ECO-DEGRADATION BIODIVERSITY AND HEALTH

B N Pandey

2024

Daya Publishing House®

A Division of

Astral International Pvt. Ltd.

New Delhi – 110 002

First Published, 2002

Reprinted, 2024

ISBN: 978-81-70352-71-6 (HB)

Published by : **Daya Publishing House®**
A Division of
Astral International Pvt. Ltd.
– ISO 9001:2015 Certified Company –
4736/23, Ansari Road, Darya Ganj
New Delhi-110 002
Ph. 011-43549197, 23278134
E-mail: info@astralint.com
Website: www.astralint.com

Printed at : **Replika Press Pvt. Ltd.**

Preface

Nature is wholly and solely responsible for survival of any organism. These days, the Nature's balance has been upset due to man's extensive interaction with nature. This has resulted in environmental degradation and has posed serious problems not only to the quality of life but also to the very existence of all living beings. Increasing rate of Urbanization, Industrialization, Mining, Deforestation and Population explosion have now reached a point of no return. Man's action with environment has resulted in land degradation, soil erosion, air pollution as well as loss of bio-diversity which has resulted in a large number of diseases. Besides, increasing rate of deforestation has adversely affected the availability of products like fruits, wood, fodder, medicinal herbs and plants. As a result, a number of diseases have now become threats to the health status of human beings.

The present book is a collection of significant articles in the field of Eco-degradation and Population Health which presents a grim picture of India's Eco-degradation and Health. It is hoped that the information presented here will awaken the authorities and preventive measure will be undertaken to prevent Eco-degradation.

I am indebted to authors who kindly accepted my invitation and have contributed valuable articles for the present book.

B.N. Pandey

Contents

1

Eco-Degradation: Its Implications and Management

Omkar and K.C. Pandey
Department of Zoology,
University of Lucknow,
Lucknow - 226 007

Abstract

The impact of man on this planet has proved to be detrimental on the environment. His mad rush towards industrialisation, urbanisation and modernisation has contributed certain hazardous effects on our ecosystem. The human population explosion has resulted in the pollution of air, water and land, along with poverty and misery. Some global issues of great concern have also been originated from the same root, such as acid rains, ozone layer depletion and global warming. The poisonous gases released as exhausts from automobiles and industries along with toxicity due to heavy metals, fertilizers and pesticides present in the biosphere have contributed certain deleterious effects on the health of human beings as well as their livestocks. The piles of hazardous and non-biodegradable solid wastes are still imposing a challenge on mankind in order to fight with them and to come forward with some eco-friendly and innovative solutions. The biodiversity in the forests has already been alarmed and struggling for its survival due to heavy deforestation, expanded agriculture and urbanisation. Moreover, the changed global climatic conditions, as a result of human activities, which have tampered the ecosystem has also laid a barrier for the biodiversity to survive. All these environmental issues, which are alarming mankind, are issues of great concern and efforts shold be taken for their management before it becomes too late.

Introduction

The impact of man upon the environment has existed ever since man himself walked on the planet earth. Whatever he has done, whether it is good or bad, has affected the environment in one way or the other. As population grows, as urbanisation and industrialisation move forward and as science and technology burgeons, the potentiality for ecological disturbance and degradation increases. The mad rush among nations over the globe for modernisation and development has jeopardised the health of man as well as the other life-forms. Progress in agriculture and industrialisation resulted into an infinite exploitation of every bit of natural resource. These activities of man has severly affected all living organisms in the biosphere. The overall misuse of nature and its resources by man had disturbed the delicate ecological balance between living and non-living components of the biosphere. These creations of unfavourable conditions have threatened the survival and prolongation of the life on earth.

Population explosion is the biggest challenge, which our country is facing, as it has crossed the limit of one hundred crore. Population, ecodegradation, poverty and health, all these are inter-related. The connection between poverty and environmental destruction can hardly be over-emphasised. The bulk percentage of the population is directly dependent on the natural resources for the basic needs of food, fuel, shelter and fodder. So, as our number is growing our needs are also increasing, resulting in the over-exploitation of the natural resources, the expense of which has produced certain serious jerks to the environment. To fulfil the needs and demands, forests are cut down leaving the biodiversity in a disarray, due to which, an ecological imbalance has been created. Some most useful and economically important plant and animal species have now been listed into endangered species (Gupta, 1987). The deforestation, population explosion and other anthropogenic activities have caused certain hazardous and deleterious implications on the environment, life forms and human health. It may include pollution of air, water and land resulting in the problems pertaining to the degradation of environment and the biota, which are the burning issues, such as global warming, ozone layer depletion, acid and alkaline rains etc. The current communication gives a detailed spectrum on environmental degradation, its impact on the biosphere along with an account of its management and conservation.

Implications of Air Pollution

Air is the primary life supporting component on the earth, which has been highly threatened due to human activities leading to its degradation. The seriousness of the air-borne problem grew with increasing population density and industrial civilisation. It may arise due to presence of solid particles, liquid droplets or gases in the air in such concentrations, which can be injurious to human, animal or plant life in the environment. It can stem from both natural and man-made sources. Natural sources gradually and silently contaminate environment without producing serious effects, *e.g.* volcano eruptions, forest fires and outgoing gases from anaerobic sediments and soil. Man-made sources are the main causative agents behind environmental degradation. Industrialisation, urbanisation and technological advancements, while heading the nation on the path of progress, have also induced many physico-chemical changes in the quality of environment resulting in its degradation, which is proving detrimental to the health of man and its livestock. The gaseous emissions in the form of automobile-exhausts, chimney wastes of industries and thermal power stations have extensively damaged our atmosphere. These gaseous exhausts contain certain harmful gases (mostly CO_2, CO, SO_2, NO, NO_2, Cl_2, Br_2 along with unburnt hydrocarbons) and metals (such as, mercury, lead, arsenic etc.), which have an adverse effect on human health (Verma, 1987b; Ray, 1992; Pachauri, 1999; Khoshoo, 1986; Kanan, 1991; Jaiswal, 1996; Gupta, 1986).

Effects of Harmful Gases

Carbon monoxide (CO) is one of the hazardous atmospheric pollutants, which is widely released along with other gases, as exhausts mainly due to incomplete combustion of sulphur containing fossil fuels in automobiles, industries, heating facilities and incinerators. It has been estimated that about 350-650 tonnes of CO is annually discharged into air (Tyagi and Mehra, 1992). People moving on the busy traffic roads and congested highways are usually affected by this notorious pollutant, specially drivers and traffic policeman. It adversely affects our respiratory system by reducing the oxygen carrying capacity of haemoglobin of blood by forming carboxyhaemoglobin. Since, haemoglobin has greater affinity for carbon monoxide than the oxygen, the formation of carboxyhaemoglobin is more than two hundred times faster than formation of oxyhaemoglobin. Therefore, its effect is more severe and fatal. Further, it may cause breathing problems, headache and irritation of mucous membrane. Its high and prolonged exposure may cause unconsciousness or even death (Sharma, 1993; Omkar, 1995).

Compounds of sulphur and nitrogen have also caused certain injurious effects on the atmosphere. The sulphur compounds present in the atmosphere as pollutants comprise of oxides, along with carbonyl sulphates, carbon disulphide and sulphates (Tyagi and Mehra, 1992). Oxides of sulphur, *e.g.* sulphur dioxide, are released from combustion of sulphur containing fossil fuel, smelting of sulphur containing ores and other industrial processes. High concentration of sulphur dioxide in the air is not only dangerous to the health of man and animals, but also detrimental to paper, leather, bricks, metals, marbles etc. Constant exposure to SO_2 may cause severe lung diseases like asthma, irritation of eyes and respiratory tract.

Photochemical Smog

Nitric oxide is responsible for several photochemical reactions in the atmosphere and involves in the formation of severe secondary pollutants like PAN (Peroxy acetyl nitrate), ozone gas etc. Photochemical smog is a resultant of the gaseous pollutants *viz.*, nitrogen oxides, hydrocarbons and oxygen, which is formed due to photochemical reactions between them. It consists of highly oxidizing chemicals like ozone, nitrogen oxides, hydrogen peroxides, organic peroxides and PAN. It has an adverse effect on plants, human health and materials and may cause eye-irritation and reduces visibility. The photochemical oxidants can enter as a part of inhaled air and impairs or interferes with respiratory process besides other processes. Serious outbreak of smog may cause asthma and bronchitis in an epidemic form (Trivedi, 1993).

Nitrogen dioxide is also a chief constituent of the photochemical smog, causing irritation of alveoli, leading to symptoms resembling emphysema, anthracosis, fibrosis and prolonged and constant exposures may even cause lung cancer. This noxious gas has a disastrous effect on the lungs of the vigorous chain smokers, as it is released while smoking. The longer the respiratory tract is exposed to the polluted atmosphere, the more serious is the histological damage as well as risk of lung tumor development. The oxides of sulphur and nitrogen released in forrn of exhausts, react with precipitation in the atmosphere, thus, resulting in the acid rain.

Acid Rain

The acid rain is an invisible threat and a burning issue for the environmentalists because of its implications. It acidifies the lakes and other fresh water reservoirs, kills fishes and other aquatic life and, thus, affect the economy, soil by inducing acidity and changing the vegetation, threatening the food and freshwater supplies, damaging forests, endangering the wildlife and other sensitive creatures, corroding structures and monuments and delustering metals. It is a global threat and advanced countries of Europe and American continent are already facing this problem, while developing countries like India are in the state of alertness. The gases, mostly oxides of sulphur, nitrogen and carbon generated by the combustion of fossil fuels, petroleum products and automobile exhausts are responsible for the acid rains. The oxidants, *viz.*, H_2O_2, O_3, HO or O_2 present in the atmosphere oxidize the oxides of sulphur, nitrogen and carbon into sulphates, nitrates and carbonates, which after hydrolysis forms their respective acids. Thus, in the process H^+ ions are liberated and culminate into the acid rain (Varma, 1992).

Alkaline Rain

In a survey conducted in Calcutta, an alkalinity in the rainwater has been observed. This alkalinity in the water ranges from pH 7.01 to pH 8.33 from the months of April to September (Mukherjee and Ray, 1992). Suspended particles present in the ambient air such as carbonates and biocarbonates of calcium and magnesium had been surmised to be the reason behind this alkaline rain. Its effect on the biosphere is still not calculated and studies are needed to ascertain its effects on life forms, buildings and the abiotic environment.

Ozone Layer Depletion

Chlorofluorocarbons (CFCs) are the chlorinated gases, which are released from the refrigerators, air conditioners, cleaning solvents, aerosol propellants etc. in huge quantities are imposing an ultimate threat on the ozone layer. This ozone layer is present in the stratosphere (12-50 km above the earth surface) and acts as a protective shield (also termed the ozone umbrella) against harmful radiations, such as ultraviolet rays along with cosmic rays and gamma rays. The exposure of this radiation to human beings may cause a significant rise in the occurrence of skin cancer along with other skin diseases and eye damage. It is also harmful to several biological species. It is estimated that each 1 per cent decline of ozone may cause 4-6 per cent increase in certain kinds of skin cancer (Warr, 1990; Dixit, 1992; Cohn, 1987).

The chemical analysis of ozone layer revealed that particular configuration of the ozone molecules allow them to absorb the ultraviolet radiations along with other radiations that are harmful to life if they penetrate to the earth surface. It has been surmised that CFCs, on reaching the stratosphere encounter with high energy ultraviolet light, which breaks them down releasing their chlorine atoms. The chlorine atoms can then engage with ozone in a catalytic reaction in which each chlorine fragment can destroy upto 1,00,000 ozone molecules before other chemical processes remove the chlorine from atmosphere (ReVelle and ReVelle, 1974). An important aspect to these CFCs is that these molecules are very stable; CFC molecules emitted today will exist to deplete the ozone for about a century. Eventually, if not a singly CFC molecule is allowed to tamper with ozone layer, still ozone losses will continue but a slower rate. Thus, crisis of the ozone layer is an important, serious and global. Attempts should be taken for its management.

The use of eco-tested and eco-friendly alternatives and substitutes against these notorious CFCs has now already been started. Hydro-fluoro-carbons are the best alternatives for the harmful CFCs such as HFC-32, HFC-134, HFC-134A, HFC-125 and HFC-143. The chemical industries in the country as well as at global level are fast gearing up to replace CFCs with HFCs. Researches to develop more eco-friendly HFCs is now pacing on the fast track. Recently, HFC-134A has successfully been synthesized at the I.I.T.C., Hyderabad (a CSIR laboratory) by a team of scientists under the leadership of Dr. A.V. Ramarao, Director of the Centre. The HFC-134A is hydro-fluoro-carbon-134A. It has no ill effect on ozone layer as well as environment (Anonymous, 1994). Prior to its discovery HFC-8, another alternative for CFCs was also synthesized which is certainly inferior in quality and performance as compared to HFC-134A.

The eco-friendly refrigerator and air conditioner had been experimentally developed, based on LPG (Liquified Petroleum Gas), in place of CFCs by a research team of Thermal Science Division, Department of Mechanical Engineering, Anna University, Chennai (Anantharaman, 1994). The LPG has a zero ozone depletion potential and its contribution in gloabl warming is also comparatively much lesser than CFCs.

The Montreal Protocol on "Substances that deplete the ozone layer", calls for fifty per cent reduction in CFCs level in time span of 15 years with a proposal to impose complete ban on its use by the year 2010. Forty-nine nations including Canada, USA and developed countries of Europe, which are together involved in 80 per cent release of CFCs have ratified the Protocol (Warr, 1990).

Global Warming

Carbon dioxide (CO_2) released in the atmosphere due to burning of fossil fuels has further placed certain disastrous attributes on the environment. Due to its increased concentration along with other green house gases (such as CH_4, CFCs, SO_2 and NO_X), a thick layer of gases is formed, which prevents the re-radiation of heat from the earth surface culminating in a rise in mean global temperature. This is referred to as *global warming* or *greenhouse effect* (Majumdar, 1999). It can lead to significant and unpredictable changes in global climate, which affects life forms including the mankind in various ways. An increase in the mean global temperature can uplift the sea level due to the melting of ice caps and glaciers which may cause high risk of flooding, associated with fiercer storms and hurricanes, thus, resulting in the loss of lives, land, biodiversity, epidemics and the poverty. The agricultural zones are expected to be severely affected by the calamity, which are in low-lying lands along the seacost. The rise in sea level may lead to millions of refugees to the world due to the shortening of coastal boundaries. Further, it would have major implications on agriculture, fishing and wildlife habitats. A number of countries such as Maldives will have to face a frightening future (Foley, 1991; White, 1990; Jones and Wigley, 1990; Hecht, 1990).

This rise in temperature also triggers widespread alteration in rainfall pattern. Precipitation is a more important determinant of survival for many species than temperature. Due to which, some regions would see dramatic increase in rainfall other may even lose their present vegetation because of drought. This change and imbalance in climate and vegetation may adversely affect the biodiversity on land. This issue was heavily debated in the recently held UNCED (United Nations Conference on Environment and Development) in Rio-de-Janeiro in June-1992 and included in the *Earth summit* (Foley, 1991). The inference drawn from this conference was that biodiversity would face major problems in the form of heavy reduction of some plant and animal species or a toal extinction of some species may occur, which could not adopt themselves to this rapid changing conditions (Foley, 1991).

Toxic Effects of Metals

Certain toxic metals, such as lead, mercury, zinc, cadmium, etc., find their way in various components of the biosphere have contributed heavily in the destruction of environment causing ill effects to plants and animals inlcuding the human beings. The ill effects include the poor health conditions, reduced growth and weight, chronic diarrhea, multifocal alopecia, excessive lacrimation and declined fertility.

Lead Poisoning

Lead compounds, which are added to petrol or gasoline, as an antiknocking agents are emitted in air as volatile lead halides along with the exhausts. About 60 per cent of these volatile lead halides remain suspended in air and when they are inhaled, they may cause reduction in haemoglobin formation leading to anaemia. Moreover, these lead compounds may also damage RBCs along with the damage to liver and kidney in man. The rest 40 per cent lead particles suspended in air are deposited on soil and vegetation near highways, which serve as the potential sources of lead to the grazing animals. Certain hazardous effects of lead poisoning are observed, which are derangement of central nervous system, gastrointestinal tract, muscular coordination and haemopoietic systems (Omkar, 1995; Pachauri, 1999).

Mercury Poisoning

It is also becoming common in animals with excessive use of mercury in industrial processes and as a fungicide. The treatment of grains with antifungal mercurial compounds is the commonest cause of mercury poisoning. Mercury related with industrial activities contributes significant amount of this metal to the environment. It has a deleterious effect on gastrointestinal and urinary systems. The acute mercury poisoning may lead to protein precipitation with coagulation necrosis in the animals (Omkar, 1995; Pachauri, 1999).

Cadmium Poisoning

It is also related with industrial and human activities. Industries engaged in extraction, refining, electroplating and welding of cadmium containing materials are major sources of cadmium in air. It is poisonous at very low levels and is known to accumulate in human liver and kidneys causing hypertension, emphysema and kidney damage. It may turns to be carcinogenic in mammals (Omkar, 1995; Pachauri, 1999).

Fluorosis

The toxic fluoride precipitates emanated by factories processing clay, glass works, phosphate refineries and aluminium factories are the main sources of air-borne fluorosis. Ruminants grazing on pastures contaminated with fluorides are the main victims of fluorosis, due to which growing animals show growth retradation. Reproductive and galactogenic performances are reduced and the animals become almost sterile (Pachauri, 1999).

Impacts of Fertilizers and Pesticides

Modern agricultural managerial systems have resulted in the increased consumption of various agrochemicals including the fertilizers and pesticides. Their production and consumption have increased remarkably with the time. Though they are beneficial to agriculture, they are also associated with the degradation of air, land and water. The pesticides, such as chlorinated hydrocarbons, organo-phosphorus compounds and carbamates, are incorporated in the air by spraying and dusting in agriculture operations. These constitute a great danger to the animals as well as man. Chlorinated hydrocarbons are used in certain persistent insecticides, such as DDT, DDE, eldrin etc., having half-lives as long as 15 years. After spraying, they incorporate themselves into atmospheric clouds and percipitate into the streams and rivers. Pesticides not only degrade the atmosphere but also contaminate the water-bodies, where they reach indirectly and unintentionally by run-off from agricultural fields, spray drift, rainwater, sewage and effluents from industries manufacturing pesticides or using them in their processes (Omkar, 1983).

WHO analysed pesticides from cereals, pulses, milk, eggs and meat samples from across the country. Of which, 50 per cent samples contained pesticide residues and in more than 30 per cent of the samples, the residue were far in excess of their tolerance limits. In India, where there is direct dependence on the above-referred food items, the human health is often severly be affected if the food contains traces of pesticides. Clearly misuse of pesticides may often lead major problems. They may enter in the human body through limbs causing beefy dermatitis. Their presence in human placenta of pregnant women has also been linked with miscarriages, stillbirths and deformities. Industrial Toxicology Research Centre (ITRC), Lucknow has strongly warned that stay or even passage of pregnant women in the pesticides sprayed areas. Majority of pesticides also affect central nervous system while some of them are carcinogenic (Singh, 1986).

In order to reduce the dependence on the pesticides, recently emphasis has been laid on adoption of integrated pest management (IPM). IPM is an ecofriendly approach to regulate the pest populations below economic injury level by employing various pest control techniques, such as cultural practices, crop management practices, biocontrol, use of pheromones and hormones, besides the rational use of ecofriendly pesticides, such as pyrethroids, biopesticides in cases of emergency, in a compatible manner with a broader objective to minimise the use of hazardous chemicals in modern agriculture systems as far as possible (Shukla *et al.*, 1998; Subhash Chander, 1999; Omkar and Srivastava, 2000; Omkar and Pervez, 2000).

Implications on the Aquatic Systems

Water is an important constituent of life supporting systems next to the air. As it is an important natural resource, we rely heavily on it for drinking, domestic purposes, irrigation, industry, shipping and for sanitation and disposal of wastes. Degradation of the water bodies (*e.g.*, ponds, lakes, rivers, estuaries and oceans) takes place mainly due to the disposal of untreated sewage, industrial effluents and agricultural discharges directly into the water reservoirs. It has produced certain serious and deleterious consquences on aquatic flora and fauna, besides deteriorating the water quality. Mortality rate of aquatic fauna is constantly increasing due to the high toxicity of the hazardous wastes in the aqua-environment.

Rivers, which are the source of the drinking water, are now gradually turning into an open sewage (Pandey, 1999; Nair, 1999).

The polluted water that contains sewage, detergents, human and animal wastes, fertilizers etc. cause eutrophication, which is large increase in standing crop phytoplankton, also known as *algal bloom*. It causes vigorous growth of algae and vascular plants in shallow water bodies. In the extreme cases of very productive hypertrophic water, there can be noxious algal blooms, an off-flavour of drinking water, the problem of toxic substances by algae and other micro-organisms, periods of hypolimnotic oxygen depletion causing mortality of fishes and other biota and evolution of noxious gases, such as hydrogen sulphide. These characteristics adversely affect the multiple use of a hypertrophic water body for the purposes of drinking water, fishery and recreation. (Pandey, 1999; ReVelle and ReVelle, 1974).

Microbial Proliferation in Water

The disposal of sewage and organic wastes composed of faeces, urine, food products, cellulose (paper) etc. has heavily contaminated the water system. Due to which, the drinking water sources such as lakes and rivers are now turning into a culture medium for the harmful microbes including certain pathogenic strains. More than hundred different viruses and bacteria are excreted in the human faeces. Many more of them are proliferated in the water bodies through other sources. The most common water borne diseases transmitted by them are diarrhea, dysentery, poliomyelitis, hepatitis, typhoid, cholera etc. Diseases like giardia enteritis, giardiasis and amoebiasis caused by protozoans, *Giardia lamblia* and *Entamoeba histolytica* are also common in the country (Verma, 1987a). Nowdays, a hike in these water-borne diseases has been observed which have spread due to toxicity and consequently increased microbial populations in water.

Contamination of Water Due to Oil

The oil used from both industrial and non-industrial sources, apparently constitute bulk contamination of water through oil. The same oil, which is for example being used for the lubrication of cars, trucks and other vehicles as well as industrial machinery. The drained oil creates nuisance for both air and water ecosystems, as certain fraction is added to the atmosphere by the combustion process and the rest may probably reach the oceans in one way or the other. Accidents involving offshore drilling operations or tankers and ship accidents also pollutes large water bodies. Bulk contribution of oil and oil products has come from refineries and petrochemical plants. There are instances in refining of oil where accidental or deliberate oil spills have occurred. Normal sea transportation by tankers and other types of ships also add tonnes of oil to the aquatic environment. When these oils and chemicals are drained off from the land to the water bodies, they disturb the natural flora and fauna, specially fishes. The effects of oil are proved to be more severe and fatal in the presence of dispersants and detergents. Since, they emulsify the oil and make it more toxic and hazardous to the aquatic biota. The emulsified oil adheres to the gill surfaces of the fishes, leading to the death of fishes by suffocation (lack of respiration). Surface feeding fishes are the main victims of oil spills.

Aquatic birds are also vulnerable to the harmful effects of oil contamination. Mostly, the diving birds are affected, which continually hit the surface of water in search of food. It has been estimated that millions of birds are killed each year due to oil pollution. Crude oil, fuel oil and waste oil are known to contain carcinogens. Excessive incidences of cancer have been reported in tar workers and men in engineering profession related to oil industries (Pandey, 1999, ReVelle and ReVelle, 1974).

Thermal Hazards

The discharge of unutilized heat from the thermal power plants directly into cooling water sources constitutes upto 70 per cent of the total heat produced in the thermal power stations (Varshney, 1983). This percentage is much higher for the nuclear power stations. The heated water reduces the dissolved oxygen content and degrades organic matters in aqua-environment including life forms. Besides fish mortality, loss in weight is also observed in fishes because of blood sucking parasites, which proliferate in water due to rise in temperature. This parasites proliferation is also a hazardous output of thermal pollution (Pandey, 1999).

Water Contamination Due to Mercury

Mercury is a potent hazardous substance, in both the organic as well as inorganic forms and is released in water reservoirs in the form of industrial discharges. It is responsible for Minimata epidemic that had caused several thousand deaths in the Minimata bay of Japan due to the consumption of heavy mercury contaminated fishes. The symptoms of Minimata disease are malaise, numbness, visual disturbance, dysphasia, ataxia, mental deterioration, convulsions and finally death. Mercury compounds, when enter into water bodies are biotransformed into methyl mercury compounds by anaerobic microbes. Reduction of photosynthesis in fresh and marine phytoplankton is also reported due to the presence of organo-mercurial fungicides. Methyl mercury compounds are highly persistent and can easily accumulate in the aquatic food chains, the major source of human intakes (ReVelle and ReVelle, 1974; Sharma, 1993).

Radioactive Hazards

The release of radioactive substances from nuclear war materials, test explosions, and nuclear power plants drastic effect the exposed biota (Ramana, 1999). These radioactive substances are hundred times more lethal than poison. It has been joinlty suggested by science and medical experts that even a smallest dose of radiation is harmful to the human health. Furthermore, there is no real medical remedy for radiation. The treatment can only ameliorate the pain and suffering. Radiation exposure may cause serious disease problems like leukemia and bone tumors, genetic damage and infant mortality. Disease such as leukemia, bone cancer, thyroid cancer and lung cancer manifest somatic effects of radiation (ReVelle and ReVelle, 1974).

Radiation damages the complex molecules, such as proteins, nucleic acids, carbohydrates and fats. It can cause the cell to produce a slightly different hormone or enzyme than it was originally programmed to produce. At the same time, such cells may proliferate to generate copies of altered cells.

If the radiation damage occurs in the germ cells (either in the sperm or ovum), it may lead to defective children and the 'error' is passed on to the succeeding generations. The exposure to radiation is known to reduce the human fertility, foetal damage within the uterus, reduced survival ability of human beings and these severe impacts may be noticed even in the future generations. A mild mutation due to radiation may express itself in human as an allergy, asthma, hypertension, slight muscular or borne defects or other genetic mistakes (ReVelle and ReVelle, 1974). Not only this, owing to its longer half life, radioisotopes are also biomagnified in the ecosystems and may lead to the various health problems.

Health Hazards Due to Noise

The noise, which is an unwanted and unpleasant sound and causing irritation and health hazards is another product of modernisation and human population explosion. It is a wrong sound at a wrong place. Initially, it was restricted to the industrial areas but the population bursts and urbanisation, with the evolution of heavy traffic, urban crowds and development and use of electric equipments and appliances have promoted the noise-nuisance in the environment. Noise levels in many urban-industralised areas are knwon to be deleterious to human health and efficiency, which effects on sense organs, cardio-vascular system, glandular and nervous system. The unpleasant sound has a vast ill effect on the human and society. It causes annoyance, headache, irritation, ill temper, loss in working efficiency, increased level of cholesterol in blood and increased blood pressure. Intolerable noise may also cause hypertension, abnormalities in foetus, peptic ulcers and may even lead to deafness (Trivedi, 1993).

Problem Due to Hazardous Wastes

The worldwide contamination of environment with toxic chemicals and hazardous wastes has become the most concerning and serious problem of mankind today. It has been estimated that there are about eight million known chemicals. Of which, ten per cent are hazardous, if used in a wrong way or accidentally released in atmosphere. The production of industrial goods involves two major environmental problems, *viz.*, disposal of waste materials, risks of industrial accidents and exposure to dangerous chemicals. Due to the overgrowing population, piles of solid waste materials and garbage are gathering in public places, which cause obstruction in a daily life. These hazardous and solid waste consist of biodegradable and non-biodegradable wastes. Some non-biodegradable wastes, which could not be recycled, have become nuisance in our environment, which include non-degradable pesticides, heavy metals, plastics, nylon, polythene etc. The use of polybags has become a curse to the life and environment. Polybags are made of non-biodegradable substances. When they are thrown away, they create havoc which is potentially eternal. Municipal reports from regions as diverse as Assam and Haryana speak of drains logged by discarded polybags, back flow of sewage pipes and disease spreading through pond like accumulation of sewage. Plastic bags have been identified as bloated receptacles of stagnant water, enough to breed mosquitoes.

Some shocking news about polybags have recently come into the limelight. The veterinary surgeons in Jaipur have removed nearly 40 kg of plastic bags from a dead cow (Chaturvedi, 1999). A very similar report has also came from Lucknow, when three cows were found dead. After postmortem, about 20-25 kg of plastic bags were recovered from the stomach of every cow. Besides this recycled polybags contain toxic dyes which may leech into the foodstuffs, endangering the human health. Thus, the plastic bags carry with them a plethora of harmful effects on human and animal health as also on the environment.

For the management of the non-biodegradable wastes, efforts have been made to recycle them. Developed countries have already started using paper bags in place of polybags and no sooner it may happen in our country too. Recently, the Central Government has taken an unexpected regressive step through its latest draft notification, Recycled Plastic Usage Rules, 1998 (Chaturvedi, 1999). The notification bans the issue of carry-bags or containers of recycled plastic for storing, carrying and packing the foodstuffs. But its implementation is still awaited.

Recent research has given rise to an alternative of plastics. These degradable plastics are of two kinds. *viz.*, photodegradable and biodegradable. Photodegradable plastics constitute light sensitive groups incorporated into the backbone of the polymers, which can be degraded by light, resulting in the breaking of structural integrity of polymers and formation of small non-degradable fragments. In contrary, biodegradable plastics and polybags can be partly or fully degraded by hydrolytic enzymes. Polyhydroxybutyrate (PHB) is completely biodegradable plastic. It is a kind of polyhydroxyalkanoate (PHA), which accumulate as inclusion of a wide variety of gram positive and gram negative bacteria (Sonia *et al.*, 1999). There are two groups of bacteria, which can synthesize PHA. One group, for example *Alcaligenes eutrophus*, produces short chain (C_3-C_5 monomers), while other, *e.g.*, *Pseudomonas olevorans*, produces medium chain (C_6-C_{14} monomers). These eco-friendly biodegradable plastics have passed all the tests and can be proved as an effective replacement for the non-biodegradable plastics (Sonia *et al.*, 1999).

Threats of Biodiversity

Disappearance of forests or its reduction in area as a result of deforestation is causing disapperance of biodiversity specially the wildlife. Biodiversity, which is facing serious threats due to over-exploitation and habitat destruction and their conservation is now-a-days a major challenge. India is recognised to be uniquely rich in gene pools and biodiversity. It is also known as one of the world's mega-diversity zones and almost all the biogeographic zones are represented in this country. It has been estimated to contain 75,000 animal species and 45,000 plant species (Sharma, 1993). This enriched treasure of India has been alarmed and endangered now. There is an urgent need for its conservation along with their natural habitats. The States as well as the Union Government has made legislations and a number of Wildlife Acts from time to time for the conservation of threatened wildlife and their habitat. *In situ* conservation of wildlife includes comprehensive system of protected areas, which consist of National Parks and Sanctuaries. Special Projects are launched for the protection and conservation of endangered species like Project Tiger for the protection of Royal Bengal Tigers. Several more projects like Project Elephant, Project Rhino, Gir-lion Project and Crocodile breeding project are turning out very effective conservation measures.

Several non-government initiatives have preceded and indeed catalyzed the reconstruction and the conservation of biodiversity. These include grassroots programmes, where small tribal and rural communities have spontaneously organized themselves to manage common lands or protect biodiversity. Chipko activism by Mr. S.L. Bahuguna in Himalayas, as well as activities of well-funded urban-based people oriented approach for the eco-restoration and eco-upgradation is beginning to take root (Gadgil, 1993; Balakrishna, 1999).

Management Measures

The ultimate problem for the eco-degradation is our ever-expanding population. It has eaten our forests and degraded our air, land and water resources. Therefore, efforts should be made to put an immediate halt to ever-increasing human population, which should be our first step towards the protection of environment.

Environmental conservation is not a new concept for us. We had a long tradition of love, respect and reverence for nature. As we are in the new millennium we have seen that people are now aware and more concerned to the environment and allied problems. Public awareness is very important as far as fight against eco-degradation is concerned. To generate public awareness, Government of India have initiated programmes on population check by celebrating the World Population Day on 11th July every year. Thanks to the Supreme Court of India for enforcing the use of unleaded and low-sulphur containing petrol for the vehicles.

As the water bodies are the source of drinking water and natural habitats of the aquatic fauna, its management is of prime concern. Restoring the water quality of our oceans, rivers and lakes is an important challenge, for which constructive strategies should be made. There are some of the most sophisticated industrial plants but still huge number of outdated industries which are constantly degrading the water systems. Various kinds of wastewater, which include industrial effluents, sewage/sludge of municipal, thermal pollutants are recycled for the beneficial use. Certain pollutants present in the water bodies can be removed by appropriate methods, such as absorption, electrodialysis, ion exchange, biodegradation etc. Government and Administration have now taken serious steps towards the purification and maintenance of certain rivers. Ganga Action Plan is one of the projects, which is giving certain promising results that may light a ray of hope for a better future. The various Legislations are made and enforced for the protection of environment and the biodiversity, but promising results have yet to come forward. Substantial resources, technical expertise, political and social will, public awareness and support are required to implement these Legislations. New technologies and equipments are needed in the industries for the treatment of water effluents before their disposal into water systems.

The global issues like ozone layer depletion and global warming are becoming horrid nightmares. The UNEP had appropriately chose the slogan "Global Warming: Global Warning" to alert and caution the people on World Environment Day, *i.e*,. on June 5 every year.

The initiative for these integrate environmental issues for consideration in development planning came not only from the Government but also from Non-Government Organisations (NGO's) and individuals. India has a very active movement by NGOs in the area of environment, large number of them involved in the rural development, afforestation programmes, creating public awareness, tribal welfare and action against pollutiong industries. Many collaborative and joint programmes exist today between the Government and NGOs. The efforts, jointly ventured by the Government and NGO collaborations, supported by public awareness may bring a new revolution and give us some fruitful results against the health hazards and eco-degradation.

It is expected from the class of people, who are seriously aware of the major consequences and implications connected with environment, to form further forceful organisations, which may jointly educate the people and suggest ways and means to the Government and related agencies to adopt eco-friendly measures and serious actions against the culprits playing a major role in the environmental destruction. Really, the destruction and degradation of the environment have turned out so vast that it cannot be fully amended, but a team effort of Legislation, Government, NGOs and public awareness can still make this planet a better place to live and to give a better tomorrow to our coming generations.

References

Anantharaman, A. (1994). Eco-friendly fridge developed. *The Hindustan Times, Lucknow,* March 2, 1994.

Anonymous (1994). A substitute of harmful CFCs, which deplete ozone layer is found. *Rashtriya Sahara, Lucknow (Hindi),* May 5, 1994.

Balakrishna, P. (1999). Biodiversity, halting steps. *The Hindu-Survey of the Environment. (Editor N. Ravi, Chennai,* 141-147.

Chaturvedi, B. (1999). Plastic bags, a modern-day blight. *The Hindu-Survey of the Environment. (Editor, N. Ravi, Chennai)*: 175-177.

Cohn, P.J. (1987). Chlorofluorocarbons and the ozone layer. Biosci. 37(9): 647-650.

Dixit, D.K. (1992). Challenges of change, ozone safe refrigerants. *Technorama,* Jul-Sept. 31-37.

Foley, G. (1991). Global warming, Who is taking the heat? Panos Publications Ltd., London pp. 104.

Gadgil, M. (1993). Biodiversity and India's degraded lands. *Ambio* 22(2-3): 167-172.,

Gupta, S.K. (1987). Consequences of deforestation. *The Hindustan Times, New Delhi*, March 7, 1987.

Gupta. Y.P. (1986). Pollution Poisoning. *The Hindustan Times, New Delhi,* November 12, 1986.

Hecht, J. (1990). Global warming: back to the future. *New Scientist*, Dec. : 38-41.

Jaiswal, K. (1996). The air we breathe is the air that kills. *The Times of India*, Lucknow, December 11, 1996.

Jones, P.D. and Wigley, T.M.L. (1990). Global warming trends. Scientific American, Aug. 66-73.

Kanan, K. (1991). *Fundamentals of Environmental Pollution.* S. Chand & Co., New Delhi.

Khoshoo, T.N. (1986). *Environmental Priorities in India and Sustainable Development.* Presidential Address, Indian Science Congress Association, New Delhi.

Maheshwari, J.K. (1993). Waste mangement: a global problem. *The Times of India, Lucknow,* November 27, 1993.

Majumdar, D. (1999). Enhanced greenhouse effect and its impact. *Everyman's Science* 34(3): 114-123.

Nair, R.M. (1999). Industrial Pollution, Belligerent attitude. *The Hindu-Survey of the Environment 1999* (*Editor N. Ravi, Chennai*): 179-183.

Omkar (1983). Effects of some insecticides on certain systems of freshwater prawn *Macrobrachium lamarrei* (H. Milne Edwards) Ph.D. Thesis. *University of Gorakhpur*, p. 257.

Omkar (1995). *Concepts of Toxicology*. Shoban Lal Nagin Chand and Co., Jalandhar.

Omkar and Pervez, A. (2000). Love fragrance in Insects. *Indian Farming* (Communicated).

Omkar and Srivastava, S. (2000). Strategies of Insect Pest Management. *Indian Farming Digest.* (Communicated).

Pachauri, S.P. (Dec. 1999). Effect of air-pollution on livestock health. *Indian Farmers' Digest.* 24-28.

Pandey, K.C. (1999). *Concepts of Indian Fisheries.* Published by Ms. Meera Pandey, Lucknow, pp. 364.

Patwardhan, A. (1999). Climate change, Shifting the burden. *The Hindu-Survey of the Environment. (Editor N. Ravi, Chennai)*: 127-133.

Ramana, M.V. (1999). Underground test, Ravaging nature. *The Hindu-Survey of the Environment. (Editor N. Ravi, Chennai)*: 7-15.

Ray, P.K. (1992). *Pollution and Health.* Wiley Eastern Limited, New Delhi.

ReVelle, C. and ReVelle P. (1974). *Sourcebook on the Environment.* Houghton Mifelin Co., Boston.

Sharma, P.D. (1993). *Environmenal Biology and Toxicology*. Rastogi Publications, Meerut.

Shukla, O.P., Omkar and Kulshreshtha, A.K. (1998). *Pesticides, Man and Biosphere*. APH Publishing Corporation, New Delhi. p. 424.

Singh, R. (1986). Pesticides residue. *The Times of India*, Lucknow, June 2, 1986.

Sonia, Jaiwal, P.K., Ahad, A. and Sastry, K.V. (1999). Biodegradable plastics: An alternative to synthetic plastics. *Everyman's Science* 34(2): 80-85.

Subash Chander (1999). Biological control of insect pests in agriculture. *Everyman's Science* 34(3): 127-131.

Trivedy, R.N. (1993). *A Textbook of Environmental Chemistry*. Anmol Publications, New Delhi. p. 410

Tyagi, O.D. and Mehra, M. (1992). *A Textbook of Environmental Chemistry*. Anmol Publications, New Delhi.

Varshney, C.K. (1983). *Water Pollution and Management*. Wiley Eastern Limited, New Delhi. p. 242.

Varma, G.S. (1992). Rain acidification in India caused by nitrates. *Atmosphera* 5: 233-239.

Verma, M.M. (1987a). Health hazards from water. *The Hindustan Times, New Delhi*, April 18, 1987. 17.

Verma, M.M. (1987b). Technology for control of pollution. *The Hindustan Times*, New Delhi, February 11, 1987.

Warr, K. (Oct., 1990). Ozone: The burden of proof. *New Scientist*. 27: 36-40.

White, R.M. (1990). The great climatic debate. *Scientific American*, July, 263(1): 18-25.

2

Eco-Degradation, Health and Well-Being and the Future

Robert J. Gregory
School of Psychology,
Massey University,
Palmerston North,
New Zealand

Abstract

A Socratic dialogue presents a parable suitable for enhancing awareness of one of the most significant problems facing humankind today. A population increase has led to eco-degradation that in turn has led almost inevitably to decline in human and habitat health. Given projected growth of human populations, the situation will get worse. Those who are aware of the situation and who have the power to act, may well make decisions and act decisively to reduce the population in what they will regard as self-defense. The timing, methods and targets will depend upon which elite decision-makers act first.

Introduction

Environmental degradation and destruction, because of ecological relationships, inevitably leads to negative social, economic and political changes. Full awareness of the multiple implications of such destruction lead to decisions, lack of or partial awareness of course, means that what may be crucial decisions are not made consciously, and what eventually happens may be determined by fate, *i.e.*, Mother Nature. But, only some people are aware of what is happening with the global environment and the on-going destruction that has taken place over the past few decades.

How can you and I make ourselves more aware, and secondly, people who are unware, more capable of understanding the global situation, and of the damage that is in place? And, more than understanding is needed, for without action, understanding is but an intellectual fantasy or a wished for dream or important words. Thus, you and I must become fully aware of the extent and types of environmental damage and destruction and we must be not only determined to provide information to others, we must act to change societies based on the information we now have. The range of actions must include managing our own personal survival, the survival of our families and networks, and strategic political, policy, and programmatic changes throughout the world. Further we must act now and in concert, not later, and we ourselves must personally take part in the actions, for others may not act in time or appropriately. Thus, an environmental ethics demands that we rigorously define, analyze, choose, and make decisions now, whether such decisions are easy, or difficult, and then act to secure a future for ourselves, our descendants, all humankind, and the other species upon our globe.

A Dialogue as Parable

Son of a Sheep Farmer: Dad, the lower paddock is flooded, worse than ever. What are we to do? We only have 1000 hectares for the farm, and this is a major problem.

Farmer: Well, son, let us think through the situation we are in. Without careful thought, we might choose the wrong direction, or the direction may be set entirely by Mother Nature, and what happens may not be something of our own choosing. Come and sit down here, with me - and have a drink. I just fixed a new batch of lemonade - here is a glass for you.

Son: Thanks, Dad. I will drink to that. Look, let us have lunch - I just made a couple of sandwiches for lunch. Have this one, and while we have lunch and a drink of lemonade, perhaps we can sort out what our future is going to be.

Okay, so of our 100 hectares of land, we are suddenly without use of one fourth, due to the heavy rains. Not only that, but the rest is likely to be less productive. No matter what we do, we cannot raise enough fodder and food to support the flock of sheep we now have.

Farmer: Yes, that is true, on the land we have. We cound rent some additional land, or buy some additional feed to make up for what we cannot grow. The prices of both land and feed have been quite high lately, but may be we could go further into debt to get more food for our livestock?

Son: Yes, but we can't realistically afford to do that, can we? Do we have the money available? Can we borrow more money from the bank? I suspect that the answer is no, we can't!

Farmer: You are right. We don't have the money and the banks won't let us borrow more. Let us look at all the implications, however, for approaching a problem piecemeal is a procedure doomed to miss the big picture. First, we must isolate what we can control and what we cannot control. For example, the rains may continue and wipe out another few hectares more this year. Or, the rains may come again next year, destroying the productivity of the land now flooded once again. We must ask why the rains were so heavy this year.

Have you heard about global warming and the greater variability of weather conditions that have been predicted as a direct result? Well, if you have, Son, then you will know that we have less and less control over the productivity of our land. Further, the ratio of carbon dioxide in the atmosphere is changing, along with more intense sunlight carrying more ultraviolet rays. The rain that falls may carry sulphur, which could be good to the soil, but nearly all forms of air pollution are harmful and degrade the long-term productivity of the soil. And, given the prevailing winds, we may even be getting radioactive materials in our rainfall. Meanwhile, we have to pay the taxes, provide the milk and meat for the community, and service the debt to the bank. You know, the problems are complex and are not likely to go away soon.

That sandwitch you fixed tastes good!

In the long run, we cannot guarantee what will happen next year, and our productive capacity has been reduced, therefore, we cannot maintain the same flock size. We must reduce our stock either in numbers, or in quality. Perhaps we could add more fertilizer to the rest of our land. Or, perhaps we can buy feed from other nearby farmers, most of them will have similar problems. Because they will have similar problems, purchases are unlikely to be a reasonable or long-term solution. True, some of them will be going out of business, and perhaps we could even get their land at cheap rates. However, we have no surplus money, no money in reserve, and no ability to raise a loan from the bank. And, they don't either so they can't take away our land. The bankers don't really know how to farm, but then again, they can always use some of their wealth to buy imported food, and still get us deeper in debt.

Son: Dad, we are in a hell of fix, as I see it.

At least, the lemonade is just right!

Farmer: Son, that is a good sandwich you fixed, thanks.

We need to isolate out the choices we have, and then act based on those choices. What if we were to reduce our livestock? Should we divide the existing food between all of them, or should we slaughter some, and if so, which ones? Should we plan for a big or a small reduction in size for the future?

Son: Well, we certainly can't feed all the stock we have. If we did try, then all would soon be in poor health, and die. The livestock that produces our livelihood won't be able to produce. I can see that the veterinarian bill will get bigger and bigger. Can we cut back on the young stock, sell the least productive, and butcher some?

Dad, you have not been sipping your drink?

Farmer: Of course, the prices we would get at this time, because so many other farmers are in a similar situations in our bio-region, would be low. But you are right, we must reduce the numbers, and yes, we could slaughter the least productive. That would include the oldest, the youngest, any with a history of sickness, and so on.

By the way, I will have my drink after I finish the sandwich you made.

Son: Does that mean that my pets, and the young animals you gave me, must be destroyed?

Farmer: Yes, I am afraid, Son, that means that your pets, and the younger livestock all would have to go. In fact, at least a third or a half of existing stock would have to go, and unfortunately, at low prices. I don't see any reasonable alternatives. The size of our flocks has to match the available pressure, grain, hay, and other foodstuffs.

Son: Can we explore alternatives? Perhaps we could plant trees, or switch to pig farming, or . . . . ?

Farmer: This land and climate is better suited to farming just as we have, than any other alternative. We have been practising sound agriculture, with the best available methods. We have a ready market and close ties to the local community. If we can't make it that way, there is no way that this farm will be productive and viable with a less tried and true product, Heavens knows, I have thought about alternatives, for the past decades I have thought about what to do. Decisions have to be made, and they are terribly hard decisions.

Son: What about my future? You know, you are getting on in years now, and I am wondering if I must go to town to work. The high unemployment there makes that unattractive though.

And, Dad, why aren't you drinking your lemonade?

Farmer: Well Son, why aren't you eating your sandwich?

But now that you raise that question about futures, when I think about your future, yes, you might have to go for work in town. The farm is not suficient to support two farmers. The question of futures is not just about the non-productive livestock.

Son: What it boils down to, Dad, is that it is you or me, isn't it, Dad?

Farmer: Yes, when you get right down to it, given the constraints we are under, only one of us, if that many, can survive.

Son, if you must know, that is why I fixed your drink and why I have not consumed any of mine. You will find that the poison I dropped in will act quickly now. You won't feel any pain. You see, age and wisdom beat out youth and enthusiasm every time, in the matter of survival of the fittest. Sorry about all that, Son. Better you than me, though.

Son: Well, Dad, you thought you beat me to it, but notice you ate your sandwich. I did not eat mine. You will find that there is so much poison put in that sandwich that it will kill you in the next five minutes, so we are even. Youth and enthusiasm are at least the equal of age and wisdom!

An Analysis

A parable of two people in dialogue is a story about two and perhaps three generations of people already here on Planet Earth. The older generation has already by its actions and its acceptance of the directions set by its leaders provided an environment and a population that can no longer survive, let alone thrive. The younger generation cannot afford, if it is to have any future, to permit the elders to have access to or to use further resources. The human population must be reduced, immediately, if not sooner. Otherwise, all will die. The parable and dialogue set the stage, and though we may be fully aware, we are not talking about and are not facing up to the challenges that are present here and now.

The issues in the parable initially apper to be about livestock numbers, but in reality, as we see by the end, the issues are about human survival. Over-population in the existing environment means that either all will die, or only some will die but in that latter case, the remaining people must consciously make the decision and act promptly and decisively to kill off the others.

Now, if you think about it, what has acutally been taking place in our modern day world? The leaders in this already doomed culture have explored the problem of over-population and limited catchment areas or resources quite thoroughly (Malthus, 1798; McMichael, 1993; Catton, 1980; Meadows, 1972; Mesarovic and Pestel, 1975; Pimentel, Harman, Pacenza, Pecarsky, and Pimentel, 1994; Cohen, 1995; Ehrlich and Ehrlich, 1990). The questions now before those leaders and the elite who are aware of the problems, concerns the timing, the methods, precisely who will act, and which targets will serve as the focus. These people are seriously concerned about their own destiny, and have the capability to remove any threat to their survival. A key question is: who among them will be first to murder the masses of people who are surplus to requirments so that a few can survive? Advantages undoubtedly accrue to those who strike first.

If no one acts, then all are destined to destruction. If a few act, they may well survive and be able to form the nucleus of what will become, a very rich and flourishing world again. In some senses, this situation is similar to that of the game, Prisoner's Dilemma. Before looking at potential and actual methods, the environmental destruction issue needs further explanation.

Environmental Destruction

Though degradation of the land is evident around the world, the range and severity of environmental problems has increased markedly. Air, water, and land pollution, loss of productive land for spurious purposes, over-population, loss of bio-diversity, loss of forests, declining fish stocks, radioactivity, global warming, increasing carbon dioxide ratios, endemic plant diseases, mono-cropping, factory farming methods, resistance to antibiotics . . . why, the list goes on and on and on. You can read endlessly about the situation on the Internet, or just browse any newspaper, watch any television news, or listen to any radio station, for just a few days. Sustainability Review as one example, provides valuable summaries of some of these problems (http://www.eeeee.net/ee02000.htm) along with doznes of other Internet based sites.

The evidence is clear to any who choose to look and examine the state of the world. The denial of a bleak reality and a stark future has gone on for a long time too but this denial can no longer be maintained. The sustainability of Planet Earth is under threat, and this generation is the one that will deal with the issues, like it or not, in the next two decades. There is no opportunity to put off for a distant future the crucial decisions that must be made. The majority of people in the existing generations are unlikely to survive to a natural death, no matter what is done over the next few years.

Most people remain unaware of these decisions that must be taken and then acted upon. Those who are aware, and who have the power and wherewithal, may well have already poisoned the remainder, that is you and I. One has only to look to see the dying taking place now, from AIDS, malnutrition and starvation, viruses and bacteria, malaria and tuberculosis, economic sanctions, convert, and overt warfare and military expenditures, poor quality and less available health care, more prisons and fewer hospitals, lack of education, and so on. How many have cancer, heart disease, and flu, for example? We have already been poisoned, and we don't even realize that we have been poisoned. One has only to look to see the propaganda, psychological persuasion, misinformation, disinformation, lies, and utter nonsense in the mass media telling us that we are healthy, happy and financially prosperous. The turth is quite different. The masses are suffering, while a few powerful and wealthy members of the elite are prosperous.

And, if you have any doubts about any parts of this scenario, then learn about the surveillance techniques already being used to track down dangerous people, materials, and ideas. Examine the coming surveillance techniques, with computers, nano-technology, robots, satellites, devices that see through clothing, retinal images, fingerprint analysis, voice recordings, and monitoring system. See what happens if you resist having your numbers and signature and picture on identifying documents that prove you are who you say you are. All it takes is a good search on the Internet, but remember, "they" will be able to track your searches.

Methods

A number of methods have traditionally been in use to reduce populations. War, and the sending of surplus young men in particular, has always been a popular way to both stimulate economies, perhaps gain additional terrain and resources, and reduce excess population. The surfeit of weapons and encouragement of violence via television, for example, brings this destructive force home to the inner cities.

Disease is another method, any by reducing health services, in particular rationing health care, and by limiting public health measures, the population can be and is being reduced. By placing public health care under extreme financial hardship, and encouraging private health care, those who can afford to pay will enable health care practitioners to profit, while reducing the numbers of people with limited resources.

Automobiles capable of high speeds, poor quality roads, and dangerous driving promoted by televised racing, readily available and inexpensive alcohol, taxing but not eliminating tobacco products, and such activities can assure large numbers of people are killed off. Poor quality diets, foods treated with strange assortments of chemicals, and experimental drugs administered through public water systems or common crops and foods, do little to protect health.

Some control over natural phenomena already exists. The United States Air Force, for example, has plans to control the weather by the year 2025. One can speculate that the recent droughts or flooding in such countries as North Korea, Honduras and other Central American States, Belgium, the United Kingdom, The Sudan, Colombia, and elsewhere may be directly or at least indirectly caused by human interventions. In a sense, global warming is due to human interventions, but hurricane study and control has long been a goal.

The list goes on, and one has only to read the papers, watch the television news, scan the Internet, or listen to the radio to find more ways in which the human population has already been harmed. What is coming in the future is open to speculation, but certainly the recent flurry of attention to possible terrorist threats, anthrax scares, chem-trails through the sky, and bio-warfare indicates that someone somewhere is thinking through these methods. When secure means of protecting part of the population have been identified and used, and when secure means of inflicting death on a large part of the population have been found, the likelihood is that some group will exercise what they will regard as their right, duty and obligation. Secure in this context means effectively hidden from public view. That is, "they" will murder hundreds of millions or billions of people, and they will term their genocidal actions as nothing more or less than self-defense. United States military activity in Southeast Asia, Iraq, Kosovo, El Salvador, Colombia, and elsewhere has been regarded as self-defence, for example. The Russian attacks in Afghanistan and Chechnaya are deemed self-defence, against terrorists.

We can read about esoteric uses of taxpayer's money with secret projects such as Haarp, Sampson, and Echelon, and wonder exactly what indeed is taking place in our world. We can read about proposals for such devices as Brilliant Pebbles, or about secret archives such as the Opal file. But because of the classification and compartmentalization of information, even the perpetrators may never learn about the true scope and scale of activities. Information tends to leak out only decades or years after secret projects are planned and conducted. The chances are that we will not be notified when decisions are made, and official sources will certainly deny that they are involved, if we should be so fortunate (or unfortunate) as to survive the likely population reduction and crash that is coming.

Questions

What do you make of the future for the world, given a scenario that:

(a) Some 245 billionaires hold 45 per cent of the world's wealth, while we know that the earth can support around 1 or at most 2 billion people (Pimentel, Harman, Pacenza, Pecarsky and Pimentel, 1994). The surplus population we have at present is destroying the capacity of the planet to sustain life by draining away the scarce resources very rapidly. That is, at least a billion people have an average income of less than one dollar per day, and they take up more food and resources than they return to the wealthy, the government, or the powerful. That makes them superfluous.

(b) These 245 people are highly aggressive, self-created, and extremely wealthy, and they are backed up and supported by about 3 million upwardly mobile millionaires.

(c) Bio-molecular technology is now or will be in place in the immediate years, to clone human beings just as well as sheep or other animals.

(d) These wealthy people can buy such technology as they need and the scientists who have expertise in developing and using such technology at what amounts to a minimal cost for the purposes they desire, not necessarily the purposes approved of by various interested groups, governments, or morally engaged agencies.

(e) Other bio-molecular technology is now in place to create viruses that can wipe out various forms of life that might offer competition, including human beings with definable genes or DNA characteristics.

(f) Most people are unaware of the issues, further, they are not suspicious, and at least right now, they are gullible.

(g) At present, a small time slot is available, with decisions to either go ahead with an attempted takeover or not of the globe. Given at least 245 persons who are billionaires and who therefore have the capability of using their wealth to achieve virtually whatever they want, and another 3 million people with at least a million dollars who could achieve quite a lot of whatever they choose, some people could act now. A significant number of governments, multi-national corporations, and other groups have the financial resources, power, and will to act as well. The various powers that be, collectively or individually, have the ability to decide and to act, and at least some of these powers that be will think long and hard and seriously about this next step. Their purpose will be not only to survive personally or in their own group, but to also spread their genes and DNA via clones without competition from the rest of us . . . . .

What Do You Think?

What would you do? What would you do if you were one of the wealthy?

Compare that with what you would do if you were one of the poor?

When Would You Act?

Why would you act? Why would you be tempted not to act? Do you really think that other won't act?

Summary and Conclusion

The human population now exceeds the carrying capacity of the planet. The world's resources are not sufficient for the current population. The biological, medical, physical, psychological, financial, and social health of the world's population is therefore at extreme risk from the overcrowding and lack of sufficient resources to satisfy existing needs. Given projections that the population is increasing rapidly, the threat to health, livelihood, and survival of all is real and significant. From these considerations, anyone who is aware and well educated can reduce that changes. That is, action to eliminate and surplus population is likely to occur in the near future. Exactly when and exactly who is extremely difficult to predict. The question remains, what methods will be used? Given past and current methods already being used, warfare, HIV and AIDs, starvation, bad weather, flue and epidemics, and supposedly terrorist attacks are given. Such methods will doubtlessly be disguised so as to prevent discovery of the attackers.

Acknowledgements

The autor appreciates a conversation with, and comments upon this article by a colleague, John Podd, Senior Lecturer, School of Psychology, Massey University, Palmerston North, New Zealand.

References

Catton, W.R. Jr. (1980). The ecological basis of revolutionary change, Urbana, IL: University of Illinois Press.

Cohen, J.E. (1995). How many people can the Earth support? New York: Norton.

Ehrlich, P.R. and Ehrlich, A.H. (1990). The population explosion, New York: Simon and Schuster.

Malthus, T.R. (1798, 1993). An essay on the principle of population, Oxford: Oxford University Press.

McMichael, A.J. (1993). Planetary overload: Global environmental change and the health of the human species, Cambridge: Cambridge University Press.

Meadows, D.H., *et al.*, (1972). The limits to growth, Washington, D.C.: Potomac Associates.

Mesarovic, M. and Pestel, E. (1975). Mankind at the turning point: The second report to the club of Rome, London: Hutchinson.

Pimentel, D., Harma, R., Pacenza, M., Pecarsky, J. and Pimentel, M. (1994). Natural resources and an optimum human population, Population and Environment: Journal of Interdisciplinary Studies, 15, 5, 347-369.

3

Environment, Disease Pattern and Health Status in a Primitive Tribe of Rajasthan

Pramod Misra and A.K. Kapoor
Department of Anthropology,
University of Delhi
Delhi - 110 007

Introduction

There is a clear connection between environment and health. 'Environmental health comprises those aspects of human health, including quality of life, that are determined by physical, biological, social and psychological factors in the environment. It also refers to the theory and practice of assessing, correcting, controlling and preventing those factors in the environment that can potentially affect adversely the health of present and future generations' [World Health Organisation (WHO), 1993].

The individual's health is strongly influenced by their environment and their life style and that these factors produce marked inequalities of health. There are three main factors which determine how healthy a person will be (Smith, 1992):

(a) Genetics
(b) Environment
(c) Life style

Genetic endowment cannot be changed but how the individual is affected by their genes will to a large degree be influenced by the way they live their life. Environment refers to aspects of a person's surrounding such as their housing, their neighbourhood, the water they drink and the air they breathe. It is not only physical factors like air, water and soil that form our environment but also the social and economic conditions within which we live. The environment is defined as "the aggregate of all external conditions and influences affecting the life and development of an organism, human behaviour or society" (Leavell and Clark, 1965). The internal and external environments comprise the total environment. Life style refers to aspects of the way people live over which they have some control such as smoking, eating and drinking. The environment has a major influence on the life style and the life style can modify the effect of environment.

Some the major influence on health are listed in Table 3.1.

Table 3.1: Some factors that influence health

Genetic make up	
Basic needs	Warmth and Shelter
	Food and Water
	Freedom from physical violence
Physical environment	Housing
	Drinking water/sewage
	Air quality
	Food quality and safety
	Roads
	Transport
	Noise
	Workplace
Life style	Smoking
	Eating
	Drinking
	Exercise
	Illicit drug use
	Safe sex practices
	Use of medical services
Environment constraints on life style	Availability of facilities (Exercise, Food choice)
	Peer pressure
	Advertising
	Cultural pressures
	Education

Health is not static but always fluctuating on a scale which ranges between optimum health and death occur due to complete lack of health. At any one time it is the result of interaction between hereditary, environmental and cultural factors of an individual. To have a clear understanding of health in its perspectives, it is required to understand how man relates his health to the environment. Living and non-living organisms in an environment are inseparable and they constantly interact with each other all the time. Thus, we may say that health can be defined as experiences of success of a man in its efforts to respond adaptively to environmental challenges. Conversely, disease may be defined as experiences of failure experienced by man in its efforts to respond adaptively to environmental challenges. Health and disease are related to biological and cultural resources of a community in a specific environment. Though health and disease are biological concepts, these are often culturally comprehended, because these have been connected with variations in socio-cultural circumstances and habit pattern.

Tribal groups of India inhabit widely varying ecological and geo-climatic conditions in different concentration throughout the country and are distinct biological isolates with characteristic cultural and socio-economic background (Basu, 1986). The health of these tribal population groups is as such a function of the interaction between socio-cultural and socio-biological practices, the genetic attributes and the environmental conditions. The widely varying prevalent health practices, use of indigenous herbal drugs, taboos and superstitions are also responsible for determining the health behaviour and health status of the tribal groups.

From an overview of the distribution pattern of tribal groups, it becomes evidently clear that they inhabit hostile terrains. Invariably these areas are highly endemic zones for malaria and other infectious diseases. Devoid of availability of safe drinking water, proper sanitation and hygiene, coupled with primitive health practices, a number of diseases breed obviously leading to higher mortality and morbidity rates.

Most of the researches were carried out on different tribes of India to collect ethnography. Mostly these studies have put emphasis in the field of religion, kinship, family, marriage and social change in various tribes of India. Unfortunately not much work has been done to understand the concept of health, disease and remedy among the Primitive Tribes. Therefore, a need arises to study the environment, disease pattern and health problems of Primitive Tribes.

The health problems need special attention in the context of tribal communities of India. It is necessary for health functionaries to have knowledge about the culture and society of tribal people. The inadequate nature of facilities in many tribal areas, lack of respect for indigenous culture, lack of proper facility and attention towards the patients are mainly responsible for non-acceptance and distrust of tribal people towards modern medicine. In fact, anthropological studies covering different facets of the society are urgently necessary to understand some the realities of health problems prevailing among the tribal population. These studies should be need based and problem solving in nature. Primitive Tribal Groups needs special attention for the same.

In the present article an attempt has been made to study the environmental aspect, disease pattern and health status among the Saharia.

The Saharias of Rajasthan

The Saharias are the only Primitive Tribe of Rajasthan and inhabit in the Shahabad and Kishanganj Tehsils of Baran Districts. According to 1991 census, the total population of Saharias in Baran District is 47,822.

Distribution

The Saharias of Rajasthan are concentrated in two tehsils of Shahabad and Kishanganj which also coincide with Panchayat Samitis. The Shahabad tehsil can be divided into two distinct divisions (1) Upreti or upland and (2) Taleti or lowland. The upreti constitutes Kishanganj Tehsil and the Western part of Shahabad Tehsil. The remaining part of Shahabad tehsil comes under the classification of taleti.

Clans

Saharia is an endogamous tribe divided into a number of clans, each clan named after some ancestor or a place to which they originally belong. The following clans are found among Saharia:

Bedgor, Bareliya, Bhilodiya, Chackrya, Chowdarya, Dediya, Devaria, Dabaraya, Dareliya, Garwar, Gogaya, Holeriya, Jarkoliya, Jaswariya, Jajwar, Khagania, Kuhar, Kanchfriya, Kheti, Khanwar, Kheliay, Kundawar, Khadeliya, Karneaniya, Khadiya, Karuriya, Kanwal, Kaluaj, Logayahi, Magriya, Navtriya, Navriya, Parenalia, Pateja, Parondiya, Rajoriya, Rewar, Raichohan, Ragpita Chohan, Rathwar, Sanoryi, Suilwar, Sapariya, Solvia and Solanki.

From above list it appears that except Bhilodia which can be directly linked with Bhils, there are at least four clans which appear to bear the names of Rajput clans like Rai Chauhan, Solanki, Devaria, Ragpita Chauhan.

Hamlet Pattern

Saharias do not have disperse pattern of living as found among the Bhils of Southern Rajasthan. They do not have even separate and scattered hamlets. However, the hamlets are slightly far off from the main villages. Each village has a cluster of Saharia families in a separate hamlet known as 'Saharana'. In the Saharana no other caste people live. In the centre of each Saharana the Saharias construct 'Umbrella type' shelter house which is called Panchayat Bungalow where all Saharias sit together to discuss, gossip and settle their disputes. Shoes are not allowed in the Panchayat Bungalow. Women are also forbidden from entering the hall. It is believed as women come from other villages, secrecy cannot be expected from them.

Marriage and Family

Saharia is an endogamous tribe but each clan in it is exogamous. They believe in totem and taboos and observe them during rituals and ceremonies of marriage. Marriage alliance between the mates of same gotra is prohibited by virtue of their affiliation to a common totem. Widow marriage or remarriage called 'Nata' is permissible among the Saharias.

The family of Saharias is a nuclear family with the youngest son staying with the parents and the elder sons moving out with their families of procreation into separate homes.

Education

In the villages where this study was carried out the literacy rate among males was found to be 3.8 per cent and among females it was only 0.9 per cent. Health-care is also a function of education. Since the large majority of villagers are illiterate inspite of the fact that each village has one primary school, they are not quite capable of having access to health care. Even if they are aware that the ailments are dangerous, they can do nothing and feel helpless.

Transport

Again, if one assumes that people have necessary money and awareness for medical treatment, communication facilities are so poor that it is difficult to have access to medical treatment. Even the post and telegraph facilities are non-existent in the interior villages. In case any epidemic breaks out there would be massive peril before the news reaches the Block Development Officer, ADM or SDO. The way to reach the news of such epidemic to relevant authorities is to go to them on foot for a substantial or by bicycle to cover the whole distance.

Drinking Water Supply

Many infectious diseases such cholera, typhoid and polio are carried in water. The organisms that cause these diseases can get into drinking water if human excreta are not properly disposed of. It is needless to emphasize the necessity of developing safe and hygienic drinking water in each and every village of the tribal area, where high infestation in almost every village results into diverse health problems, untold miser and suffering and loss of considerable manpower.

The principle source of these infections is the pollute drinking water from village ponds, rivers, hand pumps, tanks etc. On account of the rocky terrain and lack of facilities for hygienic drinking water, people have to fall back upon the rain water collected in ponds, rivers and the hand pump.

There was recent outbreak of infectious diarrhoea among children of Sandhri village. This was caused due to contaminated water. According to an article in the newspaper 'Patrika' dated 26 September, 1999, two children died in the village Sandhri due to diarrhoea and condition of 52 children was found critical. As told by the Sarpanch (headman) Mr. Ramcharan Sitara, the case of diarrhoea was reported on 25th September, 1999 by a school teacher named Amrit Lal. Two girls (named Harman Saharia, aged 7 years and Hari Charan Saharia, aged 3 years) died in the village Sandhri due to diarrhoea. Then the S.D.O. was informed immediately about the situation in Sandhri. A team of doctors from PHC Shahabad then rushed to the village Sandhri.

Housing

Warmth and Shelter are among the most basic needs of humans. People living in damp housing are much more likely to develop ashtma and respiratory infections. This is particularly true for children. Crowded housing increases the likelihood of transmission of infection between residents.

The houses of Saharias are constructed from stones and mud. The blocks of irregular stones are palced one above the other and plastered with mud. The houses have thatched roofs. The houses are generally rectangular. There are no windows in the houses. The houses of Saharias are filled with darkness even during day time. When meals are prepared the room fills with smoke and causes suffocation.

Case study findings indicate that the conditions under which a person lives have a major effect on how healthy they will be. Lali, a pregnant and nursing mother, suffering from Tuberculosis when asked about her illness replied that "living in crowded and unhygienic environmental conditions and inadequate medical and health care facilities, heavy burden of domestic chores including carrying heavy buckets of water and less interpregnancy gap cause different health problems among women".

Houses of Saharias do not have any arrangements for light and air in the villages. Poor lighting during night hours, absence of suitable exits for smoke, *i.e.*, too much smoke in crowded and ill-ventilated house may be the cause for the high incidence of eye troubles, (night blindess). Non-availability of fresh air during night causes respiratory troubles. The traditional customs, ignorance (illiteracy) and poverty, on the one hand, and indifference and reluctance of the management in these regards on the other hand are the main cuases for their living in this condition.

Sanitation and Hygiene

Sanitation among the Saharias is very poor. It has been observed that the adult male and female go to the field in morning and evening for defecation. Females go to the field in the morning, old aged persons (male and female) and children usually go for toilet purpose around their houses. The only different between audlt and child in this regard is that the child urinates around the house during the day hours but the adult do so only at night. The existing habits of defecation may be responsible for ill health of Saharias in those areas. It may be pointed out that the diseases are caused by parasites, 'Ascaris Lumbricoides' which are conveyed from a diseased person to other healthy persons by reason of contamination and it may spread through the indiscriminate defecation. The germs of cholera, typhoid and dysentry, *i.e.,* Vibrio cholera, Solmonella typhi and Entamoeba histolica are adequately present in the stool of affected men. During rainy season these germs are mixed with stream water and this contaminated water are used by the tribals and, thus, they suffer from these diseases. These parasitic infections can be controlled by providing sanitary facilities to the Saharias.

Personal hygiene is also quite poor among the Saharias. This is due to poor awareness about the importance of personal hygiene. The state of environmental and personal hygiene was found very deplorable during the fieldwork. Due to poverty and illiteracy the Saharias are unaware and ignorant about the fatality of such poor environmental sanitation.

Food Habits

A healthy diet has to supply enough energy, protein, vitamins, minerals and other essential nutrients to maintain the body and to support physical activity (Truswell, 1990). It must also supply sufficient energy for growth in children, for foetal growth in pregnant women and for lactation in breast feeding mothers. The commonest dietary problem among the Saharias is not having enough food of any kind, leading to protein energy malnutrition. This is still depressingly common in interior villages where the diet doesn't supply enough energy and it is still deficient in particular nutrients.

In the past, food habits of the Saharias were fully wild and primitive. Various forest products and flesh of different animals formed their food. They were carnivorous. Among the forest products, which the Saharias use in the form of food include mahua, chirongi, gum, honey and amla.

The grains which Saharias used to consume in the past were of inferior quality from the point of view of nutrition. Now, the Saharias eat Jawar, Bajra and Wheat. But the consumption of cereals, pulses and vegetables is very less, mainly because of low production and the cost factors.

Maternal and Child Health

Detailed interviews shoed that the Saharias do not give any special diet to the children except the breast feed. The solid food is given normally after one year or so in most cases. The implications of this observation are born out by the prevalent high rates of malnutrition (Sukha Bimari) among the Saharia children.

No special care is taken of a pregnant women among the Saharias. She takes the routine diet during the entire period pregnancy. Milk is given only if the family has milching cattle. The pregnant women works in field also.

The Saharia women know that intake of milk, dalia, green leafy vegetables is good for the growth of foetus. But due to poverty they cannot afford to eat what they should eat during pregnancy.

Due the backwardness and malnutrition, the following diseases are prevalent in the area:

(a) Respiratory diseases

(b) Veneral diseases

(c) Jaundice and Malnutritional diseases

(d) Digestive system diseases

The major casues for the above diseases are listed in Table 3.2.

Table 3.2: Showing the diseases and their causes

Sr.No.	*Diseases*	*Cause*
1.	Respiratory diseases	Due to low living standard and bad personal hygiene
2.	Veneral diseases	Lack of personal hygiene and healthy education
3.	Jaundice and Malnutritional diseases	Malnutrition and unhygienic living
4.	Digestive system diseases	Lack of safe and unhygienic water supply

Tabel 3.3 shows the list of some of the diseases found among Saharias which are caused by vitamin or mineral deficiences. Infectious disease or their illness will increase the individual's need for nutrients and may precipitate deficiencies among the tribals on a poor diet.

Table 3.3: Diseases found among the Saharias which are caused by vitamin and mineral deficiency

Sr.No.	*Nutrient*	*Disease*
1.	Energy	Protein Energy Malnutrition (PEM)
2.	Vitamin B_2	Anaemia
3.	Vitamin A	Night blindness
4.	Iron	Anaemia (usually associated with blood loss through gut)

Life Style

The word life style is used to describe those aspects of the way a person lives over which they have some choice.

Liquor

The Saharias are fond of liquor. The brew alcohol from the wild flowers and seeds of 'Mahua'. Both the men and women consume Mahua Liquor and it is deeply enconsed in tribal life, be it ceremony of child birth, marriage or festival. They generally prepare it outside the village forests.

Smoking

The Saharias are also fond of smoking. In addition to Biri (Diopyros melanoxylon), they use clay pipes called 'Sulphi', filled with tobacco.

Drinking and smoking are also responsible for ill health of the Saharias.

Traditional Health Care System

The elements of primitive medicine are still practised in many societies, both in the less developed and advanced regions, in one form or the other. The tribals in India believe that most of the diseases, specially epidemics, are caused by evil spirits. They also believe that any breach of taboo will anger the clan god and the culprit will be punished. In fact, the tribals associate the diseases to sorcery, breach of taboo, object intrusion, spirit intrusion and soul loss (Majumdar, 1958). To protect from the wrath of supernaturals, the tribals worship benevolent and protective deities and spirits. In case of diseases supposed to have been caused by evil eye, sorcery or witchcraft, they seek the help of traditional magician-cum-witch-doctor in the community.

In the tribal society, the social and cultural activities revolve around gods and spirits. They have specific gods for each and every activity of life. Likewise, they have gods for health, diseases and calamities. They worship these deities to protect them from evil spirits, diseases and epidemics. For example the Saharias worship Tejaji, a deity related with snakes. Tejaji protect them from sankes and save their life in forests.

Like other Tribals in India the Saharias, have also erected structures, though not as elaborate, conspicuous and posing as found among the non-tribal rural communities to house the images of various kinds of supernatural beings. Both as the individual as well as at the community level, the Saharias realise the responsibility of a safe placement of such images. Represented through various symbols and figures of various forms and material, they are kept at one or different places in and around the tribal settlements or in the near proximity of natural surroundings. Such religious places remain open to one and all irrespective of age, sex etc. Man's interaction with such supernaturals does not merely end in the preservation of symbols and images. Rather, it is seen that they are worshipped and appeased on specific occasions in the year. Offerings of various kinds, sacrifice, etc., are made to them on specific occasions. The will of community holds strong to keep up such values and traditions.

The Saharias believe that these supernaturals if disturbed through any gesture or forgetfulness, can cause severe condition that would prove damaging to the community health. The Saharias try to appease these supernaturals by offering prayers, worshipping them regularly and also on specific occasions like festivals, marriage ceremonies and rituals etc. Like other tribes in India, the Saharias also had, in the past, their own explanations of sickness, on the basis of which they developed their peculiar ways of curing diseases. This ethnomedical system is still in practice to a considerable extent especially among the Saharias who live in interior areas.

The Saharias believe in the presence of malevolent and benevolent powers. These supernatural powers cause sickness to individual or community or ignorance or negligence of any sort by the tribal. To calm down the wrath and anger of these supernatural powers, the tribals have discovered different magico-religious methods. By propitiation and prayer, they believe, a cordial relationship between an individual and supernatural power can be established and human sufferings can be reduced or cured. In addition to magico-religious method of treatment, the Saharias also use various natural remedies, decoction of herbs and healing lotions.

The Saharias believe that diseases and misfortunes are caused by hostile spirits, ghosts of the dead or by the use of evil spiritual powers by enemies. That a disease which is caused by evil spirits can only be cured by spiritual powers, is the main idea behind their system of treatment. Therefore, for all chronic and acute ailments, they go to their witch doctor locally called 'Janteyar'. All diseases like typhoid, tuberculosis, tetanus, pneumonia, fits, measles are considered to be caused by evil spirits and ghosts of the dead. They also believe in the hierarchy of deities and have placed some deities in senior positions. Typhoid is believed to be caused by a senior deity so it is called by Saharias as 'Bade Maharaj'. Similarly, as they believe, tuberculosis or tetanus is caused by the angry soul of forefathers.

Tetanus, also known as 'Dhanushban' is believed to be caused by evil spirits and ghosts. While working in forests, the Saharias get hurt by their axe which often causes tetanus. Pneumonia, called as 'Dibbirog' by Saharias, is considered to be caused by the attack of 'Kaccha Mashan' - a ghost of those children who died while taking milk.

For treatment of such diseases, the Saharias go to a witch doctor called 'Janteyar'. A 'Janteyar' represents a particular deity and he is believed to be the representative of that deity or ghost. When the deity appears through him, his body starts shivering rapidly and his conducts with this world breaks for a short span of time. During such moments he tells the cause of ailment and suggests remedies. For different diseases, he suggests different treatments. To a few patients he gives 'spiritual ash' purified by reciting mantras (hymns).

Restrictions on having certain foods are also imposed by Janteyar. For example, once the family members of a typhoid patient were told to burn an agarbati (incense stick) near the patient in the morning and evening and to avoid the consumption of liquor, meat and use of oil in vegetables. The socio-cultural tradition does play an important role in the context of health and treatment among the Saharias. The common beliefs, customs and practices connected with health and diseases have been found to be intimately connected with the treatment of disease. When a person goes to traditional medical practitioner he gets socio-psychological reinforcement which he cannot get from the modern medical practitioner. The social reinforcement based on the understanding of traditional beliefs and practices is often lacking and may be one of the reasons for making a modern medical practitioner unsuccessful in the tribal areas.

The Saharias have specific gods for their health and disease, for calamities and so on all these deities have their own respective departments and areas of influence, effect and control as well as nature of action. Diseases caused by wrath of deities are listed in Table 3.4.

Table 3.4: Diseases caused by wrath of deities/ghosts

Sr. No.	*Class of disease causation*	*Disease associated*	
		Local name	*English name*
1.	Wrath of gods (Devta) or goddesses (Devi)		
	(a) Badi Mata	Badi Mata	Measles
	(b) Sitla Mata	Sitla Mata	Chicken Pox
	(c) Amba Mata	-	Safe Delivery
	(d) Tejaji	-	Snake Bite
	(e) Hanuman, Jind, Thakur	-	Fever
2.	Evil spirits and ghosts of the dead	Bade Maharaj	Typhoid
		Chotte Maharaj	Malaria
		Dibbrog	Pneumonia
		Dhanushban	Tetanus
3.	Evil eye or Nazar Lagna or Dakini Lagna	Dakini Lagna or Bhoot Lagna or Upri Hawa	Fits

For some lighter ailments, the Saharias use herbal medicines which they prepare themselves with the help of various plants found in the forests. Health and treatment are very much connected with the forest ecology. The traditional health care system and treatment are based on their deep observation and understanding of nature and environment.

Attitude Towards Modern Medical System

In the past, the Saharias always used to consult spiritual doctors for treatment of sickness. But now, with the advancement, faith in the spiritual concept of sickness is on decrease. Now Saharias have faith in allopathic doctors also. During fieldwork it was found that many of the Saharias still go to a spiritual doctor for treatment of sickness. It was found that when a person falls ill initially he consults a spiritual doctor, locally called 'Janteyar'. When the treatment of 'Janteyar' did not give any relief and the disease enters its critical stage then they went to allopathic doctor. The new generation is changing with the modern times. But the Saharias of older generation still adhere to ethnomedical practices of curing diseases. It is only when the 'Janteyar' fails to cure the patient and his condition becomes bad to worse then he is taken to the allopathic doctor for treatment.

The services of qualified doctors are inaccessible to a majority of Saharias, especially in remote villages and where these are accessible the poor Saharia patients are neglected by the doctors. In such circumstances, they prefer to go directly to nearby private unqualified doctors where good treatment at low cost is available.

The Saharia women now understand the importance of the small family and a large number of them has already undergone sterilization. Their knowledge about other methods of birth control was found to be poor. The family welfare staff, is responsible for this situation. They do not go to Saharia colonies to educate them regarding health, nutrition and family planning. The desire of family planning is already there what they need is proper guidance. They were seen by Saharais only at the time of sterilization camps when they are pressurized to complete their target. Few Saharia women told that they do not want more children. But they don't know what to do. No one comes to help us. ANM visits only at the time of sterilization camp to fill their targets.

Etiology of Disease

After analysing the data it has been found that numerous causes assigned for the diseases indicate two aspects of the diseases, symptoms and etiology. While the symptoms recognised are indigestion, continuous fever, cough and cold, all remaining conditions explain the etiology of diseases. Chart 3.1 shows all these notions in the form of framework which analytically gives us a broad-outline of various types of causes attributed here.

Chart 3.1

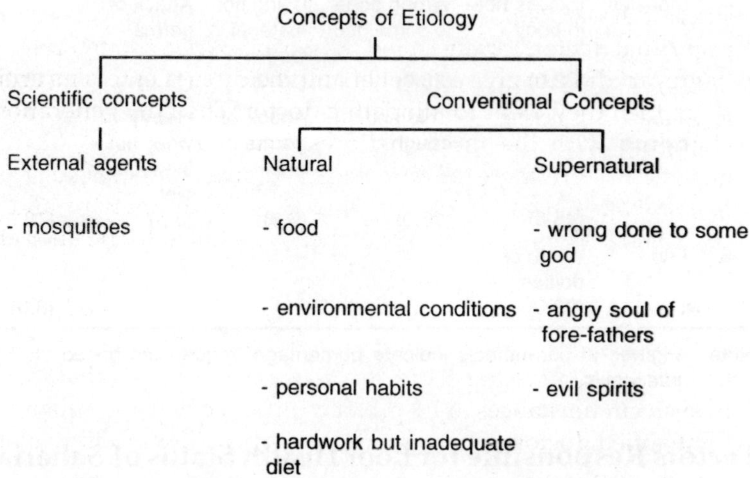

There are only two cases where the informants have indicated scientific explanations. They are mosquito-bite in case of malaria and the attack of germs in case of Tuberculosis and measles.

It has been found that the Saharias do not believe in "One cause one disease" type of thinking. Rather they believe in multi-causal explanation involving the interactions of several human and super-human factors which also include numerous socio-economic and environmental factors. Table 3.5 shows the symptoms and etiology of disease recognised by the Saharias.

Table 3.5: Symptoms and etiology of disease recognised by Saharias

Sr.No.	Disease	Causes assigned (in percentage)				Don't know
1.	Cholera	Taking rotten food (38.0)	Taking excess diet (22.0)	Indigestion (13.5)	Dirty living conditions (11.0)	(15.5)
2.	Malaria	Biting of Mosquitoes (71.0)	Dirty living conditions (19.0)			(10.0)
3.	Typhoid	Taking food during fever (70.0)	Continuous fever (10.0)	Irregularities done during fever (6.0)		(14.0)
4.	Measles	Excess heat in body (32.0)	Wrong done to some god (28.0)	Using hot material in food (19.0)	Attack of germs (11.0)	(10.0)
5.	Tuberculosis	Smoking (45.0)	Cold and cough (33.0)	Attack of germs (6.5)	Heavy work but inadequate diet (3.5)	(12.0)
6.	Fits	Wrath of deities (92.0)				(8.0)

Note: Figures in parenthesis indicate percentage frequencies based on 100 interviews.

Factors Responsible for Poor Health Status of Saharias

On the basis of our intensive and extensive field work among Saharia we found the following reasons responsible for their poor health status.

A large number of people in the tribal belt suffer from Tuberculosis. This is due to malnutrition. The food taken by Saharias is neither nutritious nor does it meet the calorie, protein and vitamin requirements of the body. They eat inferior kinds of grain like Jowar and bajra. Milk is consumed only ocassionally.

Unhygienic living conditions, delayed treatment and malnutrition affect their health badly. Even pregnant and lactating mothers and children are not provided with nutritious diet.

Poor economic condition is also responsible for low health status of Saharia. The Saharais eat inferior millets like Jowar or Bajra. Milk is consumed only by those Saharias which have milching cattle. Due to poor economic condition the Saharias are not able to buy milk and vegetables.

Several diseases are caused due to lack of safe drinking water. In villages hand pumps are not working. Pond water is used for drinking as well as for washing and bathing. Drinking water at Shahabad is having more calcium content and is heavy for drinking and requires hardwork for digesting it.

Unawareness of proper health care practices such as sanitation, hygiene, food and nutrition is also responsible for high morbidity rate among the Saharias. In the villages there were no drainage systems and sanitary arrangements.

There is no utilisation of maternity and child health (MCH) services. During pregnancy the Saharias do not consult a doctor in order to have MCH services. They approach the doctor only in case of complication or abnormality. Delivery conditions in interior villages were found to be horribly unhealthy. There is no local sweepers in some interior villages. When a delivery is conducted, she is called from other villages and it takes 2 to 3 days. This also affects mother and child's heatlh.

Superstition is also a factor responsible for poor health status of Saharia. When a person fall ill initially he resort to magico-religious treatment and if the sickness continues only then he consults an allopathic doctor.

Another factor responsible for high rate of sickness include heavy drinking habits among men which adversely affect their health. This result in malnourishment, loss of body weight and decreased resistance to infections.

Due to lack of facilities like roads and transports it is not possible for health workers to reach the interior villages to perform their duties adequately. In Rainy seaons, *i.e.,* July-Spetember it is not possible to approach villages in the interior and also not possible to conduct fieldwork especially in the villages Sandhri, Chaura Khadi and Hariyanagar. During rainy season as the area is not approachable even by jeep because the tyres of jeep get trapped in the clayey and slushy soil and that is why the village becomes unapproachable.

Lack of education is another factor responsible for poor health of Saharias. The uneducated people believe that more children means more working hands and they do not adopt family planning measures. Only women were found adopting family planning measures. They also have strong faith in magico-religious treatment of curing illness. There is no effort by the government to impart health education to the Saharias.

It was also found that some doctors and ANM are taking their duties and responsibilities lightly. They consider their posting to these tribal villages as a punishment. Some young doctors are preparing for Post Graduate (MD or MS) entrance exams. Corruption and exploitation of different nature are also rampant.

The reason for poor health status of Saharias is not only inadequate health services, but frankly speaking these are non-accessible to Saharias. There is only one Primary Health Centre (PHC) at Shahabad. Neither there is any chemist's shop in the interior villages not any doctor or health personnel to take care of poor tribals.

Suggestions to Solve the Health Problems of Saharias

Health is a function not only of medical care but of overall integrated development of social- cultural, economic, educational and political. Good health and good society go together. This would be possible only when a number of supportive services such as nutrition, improvement in environment and in education go up to a very high level. In this task special care should be taken for provision of adequate health care for the Saharias. The major health problems can be avoided by ensuring safe drinking water, curative and preventive health care, health education etc.

Great attention and importance need to be given to the protection and management of the environment. In the villages it is quite apparent that education is poor, other facilities like sanitation, drinking water, drainage, roads etc. are inadequate and the development of these facilities are not commensurate with the rural environment. Environmental education should be emphasized in the school curriculum.

Young boys and girls from the Saharia community should be given training in Public Health and Mid-Wifery respectively. They may regularly visit the 'Saharana' and advise the people on health, sanitation and the necessity of availing modern medical attention in time.

Acknowledgement

The study has been supported by a research grant from ICSSR to Pramod Misra and from University of Delhi to A.K. Kapoor.

References

Carstairs, G.M. (1955). Medicine and Faith in rural Rajasthan *(In) Health, Culture and Community*, Ed. by B.D. Paul, Published by Russel Sage, New York.

Chaudhuri, B. (1986). Medical Anthropology in India with special reference to tribal population (In) *Tribal Health: Socio-Cultural Dimensions*, Ed. by B. Chaudhuri, Inter-India Publication, New Delhi.

Foster, G.M. and Anderson, B.G. (1978). *Medical anthropology*. Published by John Wiley and Sons, New York.

Landy, David (1977). Culture, disease and healing: *Studies in Medical Anthropology*, Published by Macmillan, New York.

Gupta, S.P. (1986). Tribal concepts of Health, Disease and Remedy (In) *Tribal Health: Socio-Cultural Dimensions*, Ed. by B. Chaudhuri, Inter-India Publication, New Delhi.

Khan, Gorachand (1986). Tribal Health: Impact of Rural Development Programme (In) *Tribal Health: Socio-Cultural Dimensions*, Ed. by B. Chaudhuri, Inter-India Publication, New Delhi.

Kapoor, A.K. (1988). The Anthropologist as Biosocial hero; Health, Medicine and Faith in a scheduled tribe of Himalayas. Paper presented at the National Seminar on Health and Hygiene among the Tribal Communities in India. December 15th - 17th, 1998. Andhra University, Visakhapatnam.

Mahapatra, L.K. (1994). Concept of health among the tribal population groups of India and its socio-economic and socio-cultural correlates (In) *Tribal Health in India*, Ed. by Salil Basu, Manak Publications Pvt. Ltd.

Mann, R.S. and K. Mann (1986). Religious attributes and cure in Tribal India (In) *Tribal Health: Socio-Cultural Dimensions*, Ed. by B. Chaudhuri, Inter-India Publication, New Delhi.

Marriott, M. (1955). Western medicine in a village of Northern India (In) *Health, culture and community*, Ed. by B.D. Paul, Published by Russel Sage, New York.

Sachidananda (1978). Socio-Cultural dimensions of Tribal Health (In) *Tribal Health in India*, Ed. by S. Basu, Manak Publications, New Delhi.

Swain, S. (1994). Health, disease and health seeking behaviour of tribal people of India (In) *Tribal Health of India*, Ed. by S. Basu, Manak Publication, New Delhi.

4

Shift in Socio-Cultural Patterns by Deforestation

V.K. Sharma
Former Professor of Horticulture
3, Khera Nivas, Sanjaulli
Shimla - 171006

Since ages forests have been an integral part of our socio-cultural life, particularly for the people living in the vicinity of forests. Their day-to-day life and other activities were mainly associated with forests since they were able to get food, shelter, timber, fuel, medicines and other things of their need from the forests. Forests had all types of vegetation *i.e.* both large and deep rooted trees, bushes, climbers, shrubs and plants trailing over the ground. It formed a 3 tier system in which all the three types *i.e.* large, medium and small plants were grown.

Start of Deforestation

In the earlier days when there was less population of both man and animals, forests were the main source of food, fodder, fuel, wood, timber, local medicines etc. With the increase in population and need for more possession, man started the exploitation of natural resources near his own dwelling. Forests were one of the natural resources. He started cutting down the trees and clearing off the bushes and shrubs to make way for his residence in the first instance and then for the raising of some crops which gave rise to the agricultural practices or the cultivation of crops of his daily needs to fill his family's hunger in the first instance and then the animals and later for the future.

At the time of independence, the recorded forest of the country was reported to be about 40 m ha as per the records of Central Statistical Organisation. The government owned forest was about 26 m ha and community (ownership resting with clans, councils) and privately owned forests were about 14 m ha.

During the initial stages human beings were interested only in the present but with development of the surroundings and intelligence, he started to look for future and so adopted the habit of gathering grains and other products for later use *i.e.* when these may become scarce. Man did not stop here, being greedy in nature, started the commercial cultivaiton of crops which led to deforestation at commercial scale and the net result is before us in the form of only 23.28 per cent (in 1977) of the geographic area is under forests which includes some degraded areas.

Table 4.1 provides (Sinha, 1984) some information about the land use pattern in our country.

Table 4.1

Sl.No.	*Land use*	*Area (m/ha)*
1.	Land under forests	67.99
2.	Land under non-agricultural use like building	21.22
3.	Barren and unculturable land	19.66
4.	Uncultivated land excluding grazing land pasture, meadows	12.60
5.	Land under miscellaneous tree crops and groves not included in the net area sown	3.70
6.	Cultural wasteland	15.01
7.	Fallow land	9.59
8.	Current fallow land	13.01
9.	Net area sown	142.24
	Total	**305.02**

Natural Deforestation

Nature itself has the power to balance the vegetation. When a tree or a plant becomes too old to live or to carry on its activities, it starts withering, drying and finally dies off. The dead tree may fall off to the ground but there is no place for its fall so it causes loss to other vegetation.

Further, the climatic variations like change in temperature, speed of flow of water (streams, rivers, rivulets, gullies etc.) and wind (gails, thunderstorms, etc.), heavy rains, snowfall, movement of glaciers in the hills, helps in deforestation. Sudden change in temperature may cause splitting of earth, cracking, formation of gullies and stopping the activities of the plants and animals. Alternate freezing and thawing has a serious and worst effect on soil surface. This causes uprooting of the vegetation and its shifting to other places.

Floods and drought are responsible for large scale deforestation in hills and plains. Floods cause cutting away and denudation of rocks while drought leads to big cracks in the soil surface and drying of whole of the vegetation which cannot recover since they have crossed the limit of their recovery. Earthquake is an another form of natural hazard responsible for deforestation. Huge areas have been reported to be affected by sudden outflow of lava from volcanic rocks and gases and earthquakes throughout the world. Miles together there is no population and vegetation.

Forest fire is also a natural calamity responsible for deforestation in many areas. Such fires are natural in most of the cases. When two trees clash, particularly during the summer there is break-up of fire. On the other hand, the electric lightning may be the other cuase of forest fire. In all the cases there is a loss to vegetation which may vary with the intensity of fire.

Falling off of slips from rocks tends to cause deforestation to large extent. It has been observed that good forests are damaged by the sudden fall of slips of the mountains (due to natural or man made causes). First trees and other vegetation is uprooted and later deposited on the vegetation at another place, thus, rendering both the places deforested.

Sometimes, attack of insects-pests and diseases in the cause of deforestation. In interiors, it is not possible to control them but in areas which are near to the habitation, these can be controlled. In case of severe attack of the pathogens like locusts, grass hoppers which are most dangerous, there is severe defoliation causing the vegetation to wither, dry and die. Such losses are huge.

Artificial Deforestation

One of the most important causes of artificial deforestation is man who has deforested vast tracts of forests for his own needs. Forests have been cut for growing of crops, raising of buildings for construction uses and for other reasons like wood, charcoal, timber, fuel, etc. This has been due to pressure of population of the mankind, domestic animals and for enmassing wealth.

Changes in Life-Style

Whatever be the cause of deforestation, there had been loss of the forest wealth leading to depletion of forests, animals and birds. Many species of plants, animals and birds have become now extinct because there was a cohesion between them in the past. Trees and other vegetation provided food and shelter to the animals and birds. In return, birds, animals, insects etc. helped in the pollination and fertilization of the trees and plants; Dissemination of seeds was also carried out by them.

People living near the forests used to collect food, fodder, timber, fuel, medicines, honey, packing material, oil and other plant extracts and many other products from the forests. Apart from these some of the other activities related to these were basket making, making toy, beautification or decoration of the houses with the use of leaves, flowers and fruits of various trees and vegetation. Uses of trees are innumerable.

With decrease in area of approach of the forest dependents, life-style of the people has changed to a great extent. Some of the most prevalent changes are:

Now people have started living in isolation though they are in towns and cities but not at all concerned with the neighbour. They are involved in their own affairs and there is no community feeling.

Lack of availability of raw material from forests have forced the food gatherers to adopt some other professions. Tribals, nomads and wanders are no more visible. From the vicinity of forests they have moved to towns and cities.

The owners of land/forests have become daily wage earners since they or their parents have sold the forest at nominal price and that money has been spent and now they have to earn their livings to sustain their lives.

People have started living in houses of cement and concrete which needs lot of energy either to cool or warm them according to the season and need. Instead of tree jungles now we have concrete jungles.

In the past, people used to go to forests for sake of nature study and bird watching. Now, there is neither possibility nor eagerness to do so. Man has almost cut off his relation with the mother nature.

There may be hardly any place where you may find a tree providing shade to the passers by.

There is a loss of aesthetic sense among the people. When one is not able to see the leaves, flowers, fruits, seeds of different shapes, sizes, colours, flavour, aroma, taste, etc. how he can appreciate their beauty. The pleasure gathered by viewing the natural canopy of tree, cannot be measured in terms of money.

Losses due to Deforestation

Though technically, deforestation is removal of forest trees from the earth but in practice it has caused much loss in many ways. In a forest, hardly there is any tree which is in isolation, mostly these are grown in groups. Huge tracts of land are planted with trees.

The biggest loss is in the form of vegetation. Simply by removing a tree, its branches are lopped and some of the shoots are cut off and there remains a trunk of about 30 cm or so in the soil which is not removed by the simple means. Sometimes, this portion is left to decay or burnt. Simply removal of branches of the tree exposes the trunk of vagaries of nature *i.e.* the climatic hazards and to the attack of insects, pests. The vegetation growing under the canopy of the tree is also forced to such harsh conditions which causes them in wither, dry and finally die. These are exposed to rain, snowfall, winds (high or low), extremes of temperatures and attack of various insects, pests and diseases. Being organic in nature, the cut surface is liable to be attacked by a large number of pathogens causing the spread of diseases to the remaining vegetation.

Further, the exposed vegetation when confronted to vagaries of nature, suffers the most due to lack of shade and food material. It dwindles and dies off and whole of the ground may become open to sun and wind.

Loss of moisture from the soil surface is the cause of loss of water from the lower strata. Since there is evaporation of water from the soil surface, the water from the lower strata is forced to come up due to high temperature which necessitates the loss of water from the sub-soil also continuous evaporation of moisture causes loss in the water content of the sub-soil and even below. This may result to non-availability of water in the upper areas.

On the other hand, when there is loss of moisture from the soil surface, the activities of the flora and fauna present near the soil surface is first restricted and then stopped which renders them inactive and finally dead. On their decay, they add organic matter to the soil. They may become hosts for the pathogens to survive. This loss of flora and fauna renders the earth surface open to climatic hazards. Rain drops, when in torrential form wash away the fine soil particles along with the fine and decayed humus. Humus, the decayed organic matter is the basis of soil fertility. The surface of earth without humus becomes barren and rocky where no plant or animal may survive in the future.

Erosion of soil by water or wind and both is another loss. Soil particles and humus are carried away from the already exposed surface of the earth by washing away through water or by swaying or sweeping away through wind. Whole of the washed or sweeped material is to settle at a lower level wihch renders both the places, *i.e.* from where shifted and where it has settled down, infertile.

There is a common practice to burn the vegetation after removal of trees with the belief that there will be better growth of the vegetation in the future. This is not correct. By burning vegetation, we are lossing all the organic matter from the soil since by burning the carbohydrates and the products containing carbon are burnt and the remainder is the minerals. In the first instance, the environment is polluted by burning due to release of many gases like carbon dioxide, carbon monoxide, nitrogen oxide, sulphur dioxide etc. Secondly, the left material is mineral in nature which is in a fine state of powder that may readily be swept away or wahsed away with wind or rain as the case may be. Thus, whole of the material is lost. It also hinders the activities of the micro flora and micro fauna present in the soil. Where there is complete burning, these are completely lost.

Birds, animals, plants and human beings live in perfect harmony. All are part of our natural sustainability. Man cannot live without them and similarly others cannot. On the other hand, there is a symbiotic relationship. Birds and animals are the agents of dissemination of pollens and seeds. They are responsible for dispersal of seeds and pollens to far distant places. For cross pollinated crops these are a must. Peepal tree may not be able to grow from its own seed when planted by a man. But grows itself when its seeds are eaten by animals and birds and as such we may find such plantlets at very odd places where birds had passed their exereta. Insects, though normally harmful but are helpful in pollination of crops. Honey bees are useful for honey as they collect nectar from flowers and in doing so, they pollinate such flowers. In other flowers, insects bring about cross-pollination which may yield to new type of plants and varieties.

There may be a big change in the climate, ecology and environment. With the shift in climate as is happening globally, there is rise in temperature by about 1 to 2 degrees at present in most of the areas. Rains may be scanty or in planty but become irregular, same in the case of snowfall. Climate effects the ecology which in turn affects the environment. All these lead to eco-degradation.

At the national level, the silt removed from the hills and higher areas may be settled in plains which renders both the places unfit for cultivation of crops. Deforestation, thus, has perennial effects.

Socio-Cultural Changes

Man is a social animal. Under normal conditions, he can not live in isolation and he has to have the company, *i.e.* he must be social. Even the Adam needed company so God provided him with the Eve. Both of them roamed about in jungles where they ate apple which was a fruit from a tree. This story goes on under Christian Philosophy. According to the Hindu culture, the early man and woman were sent to this earth by Lord Vishnu and they too had to depend upon fruits of the trees and live in company. Thus, man needs company so is a social.

In earlier, days man needed food and shelter which he got from the forests. Later on, he needed some more necessities so he got shelter and then with more number of people, they started living in scattered houses or shelters. With the change of climate and necessities, his demands grew in number so he started living in villages and then in towns and cities.

Now, we do not see the thatched houses and people living in groups or communities. This is due to pressure of population on our natural resources *i.e.* water air, trees, etc. Air is beyond the control of mankind but he has polluted it by allowing emission of various poisonous and abnoxious gases into the surrounding. Water has become scarce in almost every area. Same is the condition of forests. Forests have been the hardest hit commodity since it provided the most necessary item of filling the hunger of mankind *i.e.* food. This has completely changed the life style and socio-cultural life of the mankind by way of bringing changes in peoples ideas, behavioural patterns, life-style, social systems, traditions, institutions, values and norms, expectations and other manifestations of material and non-material culture.

Since ages we have been worshipping trees, animals and birds. Our literature is full of such references and even the folk songs, folk tales and folk-lores reveal this fact. Holy Vedas, Puranas, Ramayana, Mahabharata etc. have mention of such things. Famous Indian poet, Kalidas has even described the types of trees to be raised at different places like on the boundary of a village or a town, near the house and in the kitchen garden. One can see the plantation of trees along the longest and the oldest road connecting the east of the country to the west, the Sher Shah Suri Road (previously known as Grand Trunk Road). On both sides of this road, there are trees of various types according to the agro-climatic conditions through which it passes. We generally find Mango, Acacia, Jamun, Peepal trees etc. which are all useful and good for health.

We are well aware about our customs and traditions. Tree plantation was one of the holiest celebrations. During the Gurukul system of education this was must. Teachers and taughts were involved in this. Students were to plant and maintain trees. Raising of trees was a part of their curriculum. There are reference in the medieval period which show the presence of gardens like the Yadavendra Garden, Mughal Garden, Shalimar Garden. Such types of Gardens are also in the south, the famous one is in Mysore, known as Brindaban Garden.

Worshipping of trees was a pious and sacred function, to seek blessings for a happy married life, for continuation of generations, for prosperity and well being of the community or family. We have been visiting some old trees. Such sacred trees usually of peepal, kikar, mango, banyan, bael, jamun have been part of our worship. To start a new assignment or any ritual, leaves of mango, bael, jamun are offered to the deity. There is a custom of tieing threads of yellow or red colour on branches of trees for seeking fulfilment of the wishes of the devotee.

Recenlty, a scheme for protection of the newly planted trees have been started in Delhi where one can pay a sum of Rs. 1000 per year to have a tree planted and maintained to his name. There is no doubt that this scheme is good but what is going to happen after the person who has planted and paid the money for the maintenance of the tree passes away. We have to creat awareness among ourselves, particularly the younger generation to pay respect and due regard to trees. Trees are for our benefit. They are perennials. It is a common proverb that the seeds had been planted by grandfather and grand sons are eating their fruits or enjoying the benefits of trees. It is a long-term proposal.

There is Chinese proverb. If you want to plan for a year, plant crops and if you want to plan for 10 years plant trees but if you want to plan for 100 years, educate people. So educating the people in the real sense about the use of trees has to be taken at a priority level if we want to save ourselves from the losses of deforestation. It is not the end, it is just the beginning of new era. Rise to the cause of tree awareness.

5

Foetal Wastage, Infant Mortality and Ecological and Social Factors in Six Tribal Groups of Jharkhand

B.N. Pandey, P.K.L. Das, A.K. Jha and A.K. Ojha
P.G. Department of Zoology,
Purnia College,
Purnia - 854 301, Bihar

Abstract

A survey of 1003 tribal families of the Santal, Ho, Oraon, Bhumij, Munda and Kharia of Jamshedpur belt in Jharkhand was carried out to assess the effects of industrialization and social factors on foetal and infant deaths as well as on bodily deformities. Highest percentage of foetal loss was observed in the Kharia (33.92 per cent), followed by Oraon (28.46 per cent and Ho (20.28 per cent). The percentage of infact deaths was highest in Oraon (20.92 per cent), followed by Kharia (19.65 per cent) and Bhumij (18.82 per cent). The minimum infant death was noticed in Santal (11.13 per cent) and Munda (11.39 per cent). Among the demographic factors, only age of the mother had profound effects of foetal and infant deaths. Different possible factors responsible for foetal wastage and infant mortality have been discussed.

Introduction

The Jamshedpur belt of Jharkhand is highly industrialized having a large number of factories such as TELCO, TISCO, Indian Steel and Wire Products, Indian Hume Pipe Construction Ltd., Jamshedpur Engineering and Machine Manufacturing Company Ltd., Indian Oxygen and Acetylene Company Ltd. and many others.

The area has main concentration of tribes and 93.6 per cent tribes in Jharkhand are concentrated in this area. The major tribes of the area are - Santal, Oraon, Munda, Ho, Bhumij, Kharia, Birjia, Gond, Kharwar etc. Due to industrialization, urbanization and deforestation, these tribes are facing varieties of problems.

The total tribal population is 466572 (Male - 236318 and Female - 230254). The literacy rate of male is 38.36 per cent while that of females is 13.84 per cent. But nothing is known about their foetal and infant mortality. Industrialization has also affected the biological status of these tribal communities. Keeping these facts in mind, the present work which deals with the foetal and infant mortality in some of the tribal communities of Jamshedpur belt has been taken up. The foetal and neonatal mortality rates continue to be as high as 60 - 120/1000 in India, as compared to the developed countries where it is below 10/1000 (Santhnakrishnan *et al.*, 1986; Gopaul *et al.*, 1987; Mohapatra, 1984 and Mir, 1986). Critical periods of infant mortality were observed during 1st week and 28 days after delivery (Simons *et al.*, 1978 and Mahadevan *et al.*, 1986).

Surveillance of congenital anomalies and pregnancy outcome provide important data on the effects of environmental factors. Careful evaluations of spontaneous miscarriages, malformations, prenatal, morbidity and sex-ratio all offer insights into environmental accidents (Vogel and Motulsky, 1991). Surveillance of congenital anomalies is used in many countries to provide base line data (Hook, 1990). General population rates for miscarriages are 10 - 20 per cent congenital anomalies 6 per cent and lethal defects at births 0.5 per cent (UNSCEAR, 1988).

Material and Methods

For the present work Santal, Oraons, Ho, Munda, Bhumij and Kharia inhabiting different areas of Jamshedpur belt as shown in Fig. 5.1 were selected. The area lies between 21.58 and 25 N Latitude and East Longtitude 83 and 88. The highest rainfall goes up to 150 cm and the average annual temperature varies between 45°F and 100°F. A total of 1003 (208, Santals; 195, Ho; 175, Oraons, 150, Bhumij; 100, Munda and 175, Kharia) women aged between 15 and 45 years and men aged between 20 and 55 years were interviewed, but only the female responses were used in this study. A total of 11011 pregnancies were noted (Santal, 2125; Ho, 2440; Kharia, 616; Oraons, 2090; Bhumij, 1990 and Munda, 1750).

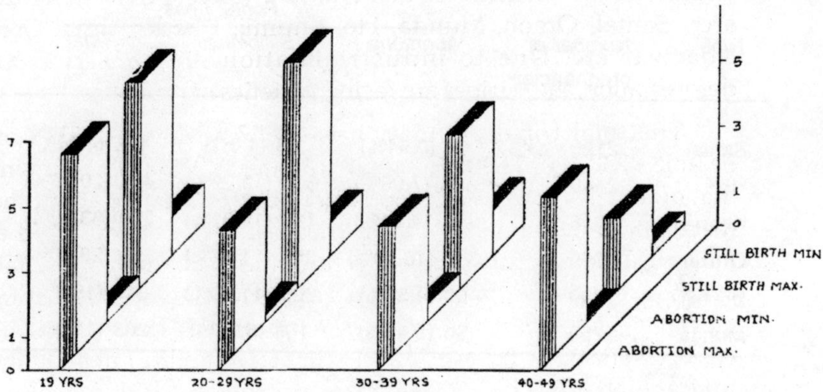

Fig. 5.1: Histogram showing abortions and still births in different age groups among tribals of E. Singhbhum (Jamshedpur)

Information about the present age of the mothers, age at marriage, the wetting period, *i.e.* gap in time between their marriage and first conception and their age at various succeeding conceptions, mortality, fertility and sex ratio etc. was collected.

Results

The percentages of gross foetal loss in the six Tribes shown in Table 5.1 and Fig. 5.1. Maximum foetal loss was observed in Kharai (33.92 per cent) followed by Oraons (28.46 per cent) and Ho (20.28 per cent). Minimum foetal loss was observed in Santals (4.94 per cent), followed by Munda (11.6 per cent) and Bhumij (15.8 per cent). The percentage of infant death (Table 5.2) was highest in Oraons (20.92 per cent), followed by Kharia (19.65 per cent) and Bhumij (18.82 per cent). The minimum infant death was noticed in Santal (11.13 per cent) and Munda (11.39 per cent).

Maximum abortions (17.54 per cent) were observed in the mother of age group 19 years and minimum (9.52 per cent) in the mother of age groups of 40+ (Table 5.4). Likewise, maximum stillbirths were observed in the mother of age groups between 20-29 years (19.8 per cent) and minimum in the mothers of age groups of 40+ (8.23 per cent).

Table 5.1: Foetal loss in six tribes of Jamshedpur, Jharkhand

Tribe	Number of pregnancies	Foetal Loss		
		Abortions	Stillbirths	Total
Santal	2125	20 (0.94%)	85 (4.0%)	105 (4.94%)
Ho	2440	195 (7.99%)	300 (12.29%)	495 (20.28%)
Kharia	616	90 (14.61%)	119 (19.31%)	209 (33.92%)
Oraon	2090	345 (16.50%)	250 (11.96%)	595 (28.46%)
Bhumij	1990	90 (4.52%)	225 (11.30%)	315 (15.8%)
Munda	1750	95 (5.42%)	108 (6.17%)	203 (11.6%)

Table 5.2: Infant mortality rate by sex in six tribes of Jamshedpur, Jharkhand

Tribe	Sex	Total birth	Neonatal deaths	Postnatal deaths	Total deaths
Santal	M	1180	62 (3.06%)	63 (3.11%)	125 (6.18%)
	F	840	55 (2.72%)	45 (2.22%)	100 (4.95%)
Ho	M	1000	69 (3.5%)	107 (5.5%)	175 (8.99%)
	F	945	60 (3.08%)	95 (4.88%)	155 (7.96%)
Kharai	M	210	32 (7.86%)	18 (4.42%)	50 (12.28%)
	F	197	18.(4.42%)	12 (2.94%)	30 (7.37%)
Oraon	M	850	109 (7.29%)	49 (3.27%)	158 (10.56%)
	F	645	105 (7.02%)	50 (3.34%)	155 (10.36%)
Bhumij	M	852	90 (5.37%)	80 (4.77%)	170 (10.14%)
	F	823	78 (4.65%)	67 (4.00%)	145 (8.65%)
Munda	M	905	68 (4.39%)	32 (2.06%)	100 (6.43%)
	F	840	40 (2.58%)	40 (2.58%)	80 (5.16%)

As far demographic factors are concerned, the maximum male child loss was observed in Kharia (12.28 per cent), followed by Oraons (10.56 per cent) and Bhumij (10.14 per cent) (Table 5.2). Maximum female child death was found in Oraons (10.36 per cent), followed by Bhumij (8.65 per cent) and Kharia (7.37 per cent). In Santals minimum female child death (4.95 per cent) was observed.

Kharia manifested maximum abortions (6.1 per cent) in the mothers of 19 years (Table 5.4) while in Santal minimum abortions (0.34 per cent) were observed in this age group. Oraon is the only tribe which showed maximum abortions in the mothers of age groups 40+ and in Santals minimum abortions were observed in this age group (Table 5.4). Maximum stillbirths (6.9 per cent) were recorded in the mothers of Kharia in the age groups 20-29 years while minimum stillbirths was shown by Santals (0.4 per cent) in the mothers of age groups 40+ years.

Table 5.3: Percentage frequency of abortions and stillbirths in six tribes of Jamshedpur, Jharkhand

Maternal age at pregnancy	*Santal*		*Ho*		*Kharia*		*Oraon*		*Bhumij*		*Munda*	
	Abortions	*Stillbirths*	*Abortions*	*Stillbirths*	*Abortions*	*Stillbirths*	*Abortions*	*Stillbirths*	*Abortions*	*Stillbirths*	*Abortions*	*Stillbirths*
19 years	0.34	1.2	2.5	2.6	6.1	6.3	4.6	3.8	1.9	4.1	2.1	1.4
20-29 years	0.20	1.0	1.9	2.8	3.8	6.9	3.2	3.7	1.4	3.5	1.4	1.9
30-39 years	0.48	1.4	2.4	4.7	2.1	3.9	3.9	2.9	1.1	2.9	1.5	1.8
40+ years	0.02	0.4	1.2	2.2	2.6	2.2	4.8	1.6	0.5	0.8	0.4	1.1
Total	1.04	4.0	8.0	12.3	14.6	19.3	16.5	12.0	4.9	11.3	5.4	6.2

Table 5.4: Abortions and stillbirths percentage in six tribal groups of Jharkhand

Maternal Age	*Abrotions (%)*			*Stillbirths (%)*		
	Maximum	*Minimum*	*Total*	*Maximum*	*Minimum*	*Total*
19 years	6.1 (Kharia)	0.34 (Santal)	17.54	6.3 (Kharia)	1.2 (Santal)	19.4
20-29 years	3.8 (Kharia)	0.2 (Santal)	11.9	6.9 (Kharia)	1.0 (Santal)	19.8
30-39 years	3.9 (Oraon)	0.5 (Santal)	11.49	4.7 (Ho)	1.4 (Santal)	17.6
40+ years	4.8 (Oraon)	0.02 (Santal)	9.52	2.2 (Kharia)	0.4 (Santal)	8.23

Discussion

Infant deaths within the first week are caused mostly by maternal factors related to delivery such as extreme age, short birth intervals, malnutrition and multiparity (Puffer and Serrano, 1975; Devanzo Butz and Habicht, 1983). The present finding shows that among the demograhic factors, only age of mothers affected the infant deaths of tribes studied. The reason for this may be that the mothers of younger age are more physiologically matured to bear a child but not physically (Madise, 1995). Higher mortality for children born to women at older ages may result from declining efficiency of reproductive system. Birth order of a child has a similar relationship with mortality as that of maternal age. Certainly among the Santals of Dalbhumgarh there is a gap period between two children for at least three to five years, thereby reducing the infant mortality. It has been reported that the higher mortality rate during first week of life may be due to *Tetanus neonaterum* - a result of unscientific way of cutting the umbilical cord with a household knife or an unsterilized blade or sickle etc. Such practices among the tribals are still common.

The nutritional status of the mother has also profound impact on mortality. Even during intra-uterine life, in the unborn state, maternal malnutrition can adversely affect the health of embryo and ultimate outcome of pregnancy. The maternal nutrition provides necessary nutrients for growth and development of foetus, the genetic profile gives direction of the utilization of these nutrients in relation to growth and development. According to Winik-off (1978), due to chronic malnutrition in expectant mothers the risk of deaths of foetus as well as mothers may be increased because lesser nutrients are available for the foetus and maintenance of health of the mothers. It increases the frequency of birth injuries and reduces the overall efficiencies of the reproductive processes. The loss of foetus during latter stage in the present study might be influenced by maternal nutritional status. The chief food of these tribals is rice. They do not consume milk and leafy vegetables. Thus, their diet is deficient in protein and vitamins. Apparently this deficiency in diet might leads to abortions.

Infant death during postnatal period might be due to various problems such as inadequate supply of food, contact with sick and unhealthy person as well as unhygienic environmental conditions.

Certainly living conditions of the Ho, Kharia and Bhumij are very poor and unhygienic. Acute diarrhoeal diseases are one the major causes of morbidity and mortality in India especially among infants and children below 5 years of age. About 1.5 million children under age of five years die of diarrhoeal disease in India every year. As living conditions of these tribes as poor and sources of drinking water are not hygienic, the infant as well as child mortality due to diarrhoeal diseases is very common among them. The principal causes of infant and child mortality are infectious, parasitic, nutritional, respiratory and other prenatal diseases. Especially in the developing countries, the high infant mortality is characterized by the predominance of infectious and parasitic diseases (Boulanger, 1980; Chajhacka and Adeghola, 1984). Further, mother eduction may reduce infant mortality by hygienic practices and survival (Wave, 1984).

Successive developmental stages of a foetus and its metabolic processes controlled and influenced to a greater or lesser extent by genes (Wright, 1970b). Any mutation in these genes do not permit the load bearing embryo to pass successfully through the 'phenocritic' period of embryogenesis. Mutational load is sometimes so severe that the individual cannot survive and death may occur very early in embryonic development leading to abortion. The high early foetal mortality recorded in the Kharai might be due to genetic factors. Evidences of direct effect of genetic factors are lacking except for the role of chromosomal anomalies (Warburton and Preser, 1964). 40 per cent of our health problems are of genetic origin (Lederberg, 1971). Further more, it has been found that the marriage distance in the present population is very short, hence chances of consanguinity/inbreeding are high (Ansari and Sinha, 1978), leading to increased genetic load among them. A gradually rising incidence of prealnatal deaths with degree of consanguinity can be directly attributed to the appearance of homozygosity at different gene loci and lethal effects of such abnormalities, resulting in abortions and stillbirths (Mian and Mushtaq; 1994 and Pandey *et al.*, 1997). Abortions may also occur due to abnormal uterine activity and death of foetus (Stevenson and Warnock, 1955), various maternal foetal incompatibilities and hormonal imbalances (Levine, 1958).

It has been reported that malaria can induce pregnancy wastage and may affect fertility by producing lesions in genital tract (Taylor,

1991). Of course malaria is prevalent among the tribal inhabited areas, particularly those by the Ho and Kharia. Increased abortion rates have been found in women in jobs requiring heavy lifting, standing long hours or exposures to excessive noise or vibration (Headapohl and La Dou, 1993). It is to be noted that majority of female tribals work as daily labourers in nearby factories or agricultural fields. Thus, their occupation may have risk of spontaneous abortions.

It has been reported that 1/200 live born babies have a gross abnormality of chromosome number of structure. High as this figure is, it does not indicate fully the frequency with which chromosomal abnormalities occur at fertilization, for in a study of early spontaneous abortions chromosomal aberrations occurred in over 50 per cent cases. Many of these chromosomal anomalies are rarely found in live-born babies and are therefore, presumably lethal to and a cause of significant number of early abortions (Emery, 1987). High abortion rates among tribals residing near UCIL complex may be attributed to chromosomal anomalies which have been taking place due to radiations.

Grosche *et al.* (1997) have reported increased stillbirths in Southern Bavaria during the first 2 years following the Chernobyl disaster. High rate of prenatal mortality was observed in Sweden during June and July, 1986 (Kallen, 1988). Buzhievsraja *et al.* (1995) have reported 9-15/100 spontaneous miscarriages in Kiev for Chernobyl effect on pregnancy. According to the data collected in Hiroshima and Nagasaki during the post 40 years on the children of the bombings, no statistically significant genetic effect emerges with respect to eight different indicators (Neel *et al.*, 1990). But unlike the inhabitants of Jadugoda the Japanese were exposed to a short intense burst of radiation.

Damage to the foetus by ionizing radiation is a special case of somatic damage. The developing foetus is particularly susceptible to damage during the period of organogenesis. When the loss of reproductive capacity of a cell rest due to evolution into a particular organ may be catastrophic. The foetus like adult, is also subject to an increased risk of the development of Leukaemia among tribal children may be due to ionizing radiation of uranium.

The period of development from the 9th day to the 16th week after conception is known as organogenesis. While organs and limbs are being formed, the embryo is most vulnerable to disease, to drugs and also to radiation. Modest amount of radiation can also have catastrophic consequences during this time. The whole spectrum of defects can also occur including cleft plate, stunting of the limbs, abnormally developed brain and so on. Growth retardation and neonatal death can also result from irradiation at this time, *i.e.*, death at or about the time of birth. It also leads to live birth of a grossly deformed child. Exposure of the developing embryo to ionizing radiation may produce growth retardation, death and / or congenital malformation. Irradiation during early organogenesis and even more so during late organogenesis, produces growth retardation stillbirth as well as loss of organs. The predominant types of malformations observed due to radiations are the central nervous system (microcephaly mental retardation), sense organs and skeletal system and testicular atrophy. On the basis of these aspects it can be said that defects in children of the Oraons inhabiting villages near UCIL might be due to ground radiation and radiation from waste disposal of UCIL. However, causes other than radiation might have influenced the results. A case control study would be necessary to obtain further information.

The bone marrow is a critical tissue, most sensitive to the effect of radiation. With increasing dose, gastrointestinal tract, skin, lungs and other organs are also affected. Total body irradiation rapidly produces profound lymphocytopenia and immunosuppression that lead to increase instances of abortions (Iyer, 1996). It is to be noted that human milk is superior nutritionally and immunologically formulas or cows milk. Breast milk is especially in developing countries where poor sanitation and lack of clean water supply further deteriorate the health of children. The World Health Organisation (WHO) recommends that even HIV infected mothers should breast feed. It is to be noted that although few infants accompany parents to work place but they can receive significant exposure to toxic chemicals through the working mother's breast milk.

Matigoda hill acts as a barrier between Kulamada and UCIL (Fig. 5.2). Kulamada is a cluster of many small 'tollas' like Beramdera, Jobla, Bakai, Borakanta, Chiruda, Dabhkucha etc. It is surrounded by hill from three sides and is rich in vegetation. The area is mainly inhabited by the Ho. Quite interestingly, in this area we noticed no cases of genetical abnormalities like leukeamia, thalassaemia etc.

but tribals inhabiting the villages nearby UCIL have a large number of such genetical abnormalities. It indicates that the Matigoda hill, having rich vegetation, protects the inhabitants of Kulamada from the harmful effects of radioactive rays. Further, it was reported by villagers that children playing in slime dam of UCIL, Jadugoda develop a large number of abnormalities.

It has been reported that alcoholic women have high spontaneous abortion rates (Abel, 1977). Thus, it can be concluded that high rate of abortion in Kharia may be due to alcoholic addiction of women. In Kharia females are highly addicted to alcohol.

Acknowledgement

The financial help for this work was provided by Ministry of Environment and Forests, Government of India, under the project entitled "Pattern of Human Settlement in and Around the North Gangetic Belt of Bihar with particular reference to certain Scheduled Tribes", under MAB Programme Ref. No. 14/5/94 Re-MAB under the guidance of Dr. B.N. Pandey (Principal Investigator, Department of Zoology, Purnia College, Purnia-854 301), is gratefully acknowledged. Authors wish to thank, Dr. M. Thomas, Professor and Head, Department of Anatomy and Division of Human Genetics, Medical College, St. John's National Academy of Health Sciences, Bangalore (India), for her valuable suggestions.

References

Abel, E.L. (1997). "Maternal Alcohol Consumption and Spontaneous Abortion". Alcohol and alcoholism, Vol. 32, No. 3, 211-219.

Ansari, N.A. and Sinha, S.P. (1978). Survey on the effects of inbreeding in two populations of Bihar. Indian Journal of Medical Research. 68: 295-299.

Boulanger, P.M. (1980). More trends in the fight against infant mortality. In: La Moralite des enfants dans le mande et. dans 1'historie. P.M. Boulanger and D. Tabulin (Eds.) Ordinia Edition.

Buzhievsraya, I., Tchaikovskya, I., Demidova, G. and Koblyanskaya, G. (1995). Selective monitoring for a Chernobyl effect on pregnancy out come in Kiev, 1969-89. Human Biology. 67(4): 657-672.

Chojhacka, H. and Adeghola, O. (1984). The determinants of infant and child morbidity/mortality in Lagos, Nigeria. Social Science Medicine, 19: 799-807.

De Vanzo, J., Butz, W.P. and Habicht, J.P. (1983). How biological influences on mortality of Malaysia vary during the first year of life. Population Studies. 44: 489.

Emery, A.E.H. (1983). Elements of Medical Genetics. 6th Ed. Churchil Living Stone. Edinburgh.

Gopaul, S., Kumar, P., Sivaraman, R. and Shanthana Krishnan, B.R. (1987). Perinatal mortality in hospitalized deliveries. In : Current trends in Perinatal Medicine. H.A. Mir and M. Hassan (Eds.) Vishinath Press, Srinagar.

Grosche, B., Irl, C., Schoetzau, A. and Van Santen, E. (1997). Perinatal mortality in Bavaria, Germany, after the Chernobyl reactor accident. Radiation Environment and Biophysics. 36: 129-136.

Hook, E.B. (1990). Timely monthly surveillance of birth prevalence rates of congenital malformations and genetic disorders ascertained by registries or other systematic data bases. Teratology 41: 177-184.

Iyer, G.K. (1996). Hundred years: ionising Radiation and Health. Guest Article - L&T Oration 46th National Conference of A.O.H. at Pune, February 1996.

Kallen, B. (1988). Pregnancy outcome in Sweden after Chernobyl. A study with central health registries. National Board of Health and Welfare, Stockholm.

La Dour, J. (1993). International issues. In women workers, D. Headapohl (Ed.). Occupational Medicine: State-of-the-Art-Reviews. Hanley and Belfus, Inc. Philadelphia.

Lederberg, J. (1971). In Forward of Mutagenicity of Pesticides. S.S. Epstein and M.S. Legator (Eds.) MIT Press.

Levine, P. (1958). The influence of the ABO system on Rh-haemolytic diseases. Human Biology. 30: 14-28.

Madise, N.J. (1995). Birth spacing in Malawi and its impact on under Five Mortality. Ph.D. Thesis, Department of Social Statistics, University of Southampton.

Mahadevan, K. Reddy, P.R., Murthy, M.S.R., Reddy, P.J., Gowri, V. and Siva Raju, S. (1996). Culture nutrition and infant and childhood mortality. In: Fertility and Mortality: Theory, Methodology and Empirical Issues. K. Mahadevan (Ed.). Sage Pub., New Delhi. Pp. 314-337.

Mian, A. and Musthaq, R. (1994). Consanguinity in population of Quetta (Pakistan). Journal of Human Ecology. 5: 49-53.

Mir, N.A. (1986). Regionalization of perinatal care and its effects on perinatal mortality. Indian Journal of Pediatrics. 53: 335-338.

Mohapatra, S.S. (1984). Rural Perinatal mortality. Indian Journal Pediatrics. 21: 645-649.

Neel, J.V. and Schull, W.J. (1990). The children of parents exposed to atomic bombs: Estimates of genetic doubling dose of radiation for humans. American Journal of Human Genetics. 46: 1053-1072.

Pandey, B.N., Jha, A.K. and Sinha, M.K. (1997). Study of blood groups in relation to reproductive performance among Muslims in Purnia Division, Bihar, Journal of Human Ecology. 8(4): 233-236.

Puffer, R.R. and Serrano, C.V. (1975). Birth Weight, Maternal Age and Birth Order: Three important Departments of Infant Mortality. Scientific Pub. No. 294. Pan American Health Organisation. Washington, DC.

Santhanakrishnan, B.R., Gopaul, S. and Jayam, S. (1986). Perinatal mortality in referral hospital in Madras City. Indian Journal of Pediatrics. 53: 359-363.

Simons, G.B., Smuker, C.M., Mishra, B.D. and Majumdar P. (1978). Pattern and causes of infant mortality in rural Uttar Pradesh. Tropical Pediatrics Environment and Child Health, 27: 207-216.

Stevenson, A.C. and Warnock, K.H.A. (1955). Observation on the results of pregnancies in women resident in Belfast. Annals of Human Genetics. 23: 382-394.

Taylor, G.C.N.M. (1991). Human adaptation and the impact of diseases. Journal of Indian Anthropological Society., 26(122): 37-49.

UNSCEAR (1988). Ionizing Radiation: Sources and Biological Effects: Report to the General Assembly with Annexes. UNSCEAR, New York.

Vogel, F. and Motulsky, A.G (1991). Human Genetics. Mir Edit., Moscow (in Russian).

Warburton, D. and Preser, F.C. (1964). Spontaneous abortion risk in man. Data from reproductive histories collected in a medical genetic unit. American Journal of Human Genetics. 16: 1-25.

Wave, H. (1984). Effects of maternal education women's roles and child care on child mortality. In Child Survival: Startegies for Research. W.H. Mosley and L.C. Chen. Supplement to population and Development Review, 10: 25-45.

Winik-off, B. (1978). Nutrition, population and health: Some implications for policy. The Population Council Working Paper, 3: 1-32.

Wright, T.R.F. (1970). The genetic embryogenesis. In: Drosophila: 1 per cent Advance in Genetics. E.W. Caspavi (Ed.). Academic Press, New York. 261-395.

6

Modernization and Demographic Change: A Study in a Primitive Tribe of Madhya Pradesh

G.D. Pandey and R.S. Tiwari
Regional Medical Research Centre for Tribals (ICMR), Jabalpur - 482003

Abstract

The study has indicated that as regards, the socio-economic characteristics, there has been not much improvement in literacy and income but the occupation pattern of the primitive Bharias is different than the non-primitive Bharias. There are 7 per cent people engaged in service among the non-primitive Bharias, compared to none among the primitive Bharias. Among the others, the longevity is higher, the dependency ratio is lower among the non-primitive Bharias. Frequency of consanguineous marriages are lower and mean marital distance is lower which are positive improvements and shows association with modernization. There is no difference in fertility indicators but the primitive Bharias have higher infant mortality rate and higher abortion rate. The non-primitive Bharia couples have adopted sterilization for family planning in higher proportions, though the age of wife at the time of sterilization do not differ considerably.

It was observed that the non-primitive Bharias are more aware about the PHC services and they use the services more frequently whenever they needed. Due to tribal transformation the non-primitive Bharias have adopted a number of practices which are not prevailing in the primitive Bharias, for example, use of medicine for the problems of lactation, consultation with the doctor in problems of delivery, postive attitude towards antenatal and postnatal care etc. but still they require much more interaction/awareness. A lot of efforts are required for upliftment of the primitive Bharias which are lagging behind with their co-brothers (Non-primtive Bharias) who are more exposed to the modernization.

Introduction

Among all the states of India, Madhya Pradesh has the largest population of tribals, about one fourth of India's total tribal population. There are as many as 46 different tribal groups in the state among which Gonds, Bhils, Oraons, Bhilalas, Kols, Kanwars are prominent. These groups show considerable variation in population size, level of development, acculturation and assimilation. Their social and cultural traits, religious beliefs and norms as well as traditional economic pursists vary from tribe to tribe (Pandey and Saxena, 1988). Out of these 46 tribal groups, seven have been identified as the most backward or primitive tribes. They are Abujhmarias, Baigas, Hill Korwas, Bharias, Kamars Saharias and Birhors inhabited in the different parts of the State (Tiwari, 1984).

This article attempts to see the effect of primitivenes on the demographic characterstics of the Bharias of Patalkot valley.

Data and Methodology

The Tribe

Bharias are one of the smaller tribal groups of Madhya Pradesh. They prefer to call themselves Bhumia. It is a drawidian tribe residing principally in the central region of Madhya Pradesh. A small section of this community, lives in an area known as Patalkot in Chhindwada district. This area is a deep depression in the hilly region, the approach being extremely difficult. This group has been bypassed by modern development because of physical situations of these habitat. The Bharias living inside valley has been accorded status of Primitve Tribe by the Government.

The Area

Tiwari (1984), Dixit (1984), Kurup (1986), Pandey (1999), Abbad (2000) have given a detailed description of the Patalkot valley. It is bowl shaped formation situated on the Satpura plateau and is located towards the north of headquarters and about 23 kms away from block and ITDP headquarters. The area of Patalkot lies between 22°-24° and 22°-29° North latitude and 78°-43° and 78°-50° East longitude. The shape of the area is like a horse shoe surrounded on three sides by hilly ridges and open on the North-west side towards Dudhi river. The area consisted of ridges and valleys. The ridges going down to the level of Dudhi river valley.

Physically the area can be divided into two natural divisions *i.e.* the hilly ridges and the valley. The highest contours of the area being at 3750 fts or about 1209 m. The hilly ridges inside the area have then altitude ranging between 887-1050 m above mean sea level. The outer ridge is like a straight wall and is locally known as "Kanat". The villages identified in the Patalkot lay within the fold of this "Kanat". The lowest contours, noticeable at the North-West portions of the area near river Dudhi are at about 485 m. The Patalkot valley has an altitude ranging between 485-1050 m above mean sea level.

There are many seasonal and perrenial streams and rivulets within the area. The difficult terrain, hard life and almost negligible development has given rise to smaller sizes of the villages. As per 1991 census, out of 12 villages, 6 villages has population less than 200 persons and another 3 had population less than 50 persons.

The Regional Medical Research Centre for Tribals, Jabalpur has collected information on socio-demographic characteristics of the primitive Bharia tribe during 1995-97. The data was collected with the help of questionnaire and personal discussion. For the sake of comparison and to see the effect of modernization on the demographic characteristics Bharias living outside of the Patalkot valley (non-primitive Bharias) have been studied. This artilce is based on the data collected in Bharia tribe living inside (Primitive Bharias) and outside (Non-primitive Bharias) of Patalkot valley. The samples of individuals studied in these two groups were 699 and 1222. The details of the observations have been discussed in Pandey (1999).

Results and Discussion

Level of Backwardness

Compared to the non-primitive Bharias, the primitive Bharias prefer to live in joint house holds. A higher percentage of population is engaged in agriculture and collection of forest produce while no one has found having occupation as service. The literacy percentage is slightly lower but does not differ significantly. Despite of the variation in occupation pattern, there is no considerable variation between annual income of the households of the two groups (Table 6.1)

Table 6.1: Comparison of socio-economic indicators of primitive and non-primitive Bharias

Sl.No.	Indicators	Primitive	Non-primitive
1.	Type of hosuehold (%)		
	Nuclear	50.3	66.9*
	Joint	49.7	33.1
2.	Electrified houses (%)	61.5	48.6*
3.	Occupation (%)		
	Agriculture	39.6	43.9
	Collection of forest product	37.4	0.0
	Labour	23.0	47.5
	Service	0.0	7.6
4.	Land/HH (Acres)	5.2	5.4
5.	Literacy (%)		
	Male	34.1	38.3
	Female	8.9	11.9
	T	21.5	25.1
6.	Annual Income/HH (Rs.)	6317	6321

Differentials in Demographic Characteristics

As evident from the Table 6.2 , it appears that non-primitive Bharias have higher proportion of aged persons (60+) and low proportion of child populations (0-14 years). The young dependency ratio which is the ratio of child population (0-14) to the working population (15-59) expressed in percentage is slightly higher among the primitive Bharias. The old dependency ratio which is the ratio of aged population (60+ years) to the working population expressed in percentage is slightly lower for primitive Bharias. Thus, 86 per cent of the Primitive Bharia population is dependent population compared to the 84 per cent among the non-Bharia populations. Similarly index of ageing which is an indicator of longevity is lower (15.5 per cent) among the Primitive Bharias than the non-primitive Bharias (18.2 per cent) (Table 6.2).

Table 6.2: Comparison of demographic indicators of primitive and non-primitive Bharia tribe

Sl.No.	*Characteristics*	*Primitive Bharia (N = 699)*	*Non-primitive Bharia (N = 1222)*
1.	Age composition (%)		
	0-14	40.0	38.7
	15-59	53.8	53.3
	60+	6.2	7.0
2.	Young dependency ratio (%)	75.5	71.3
3.	Old dependency ratio (%)	11.5	12.9
4.	Total dependency ratio (%)	86.0	84.2
5.	Index of aging (%)	15.5	18.2
6.	Sex ratio (F/M x 1000)	1068.0	1009.8
7.	Age at marriage (years)		
	Male	18.1	18.4
	Female	15.8	15.6
8.	Consaguinity (%)	55.3	25.5
9.	Mean marital distance (kms)	10.2	8.3

The sex ratio, in primitive Bharias is higher. Both groups of Bharias do not differ in female age at marriage, a factor highly associated with fertility. It was observed that among the primitive Bharias, more than half (55.3 per cent) of the marriages were consanguineous while such marriages were only 25.5 per cent in the non-primitive Bharias. Similarly village endogamy *i.e.* marriages within the village was higher among the primitive Bharias (25.3 per cent) compared to the non-primitive Bharias (13.2 per cent). The differences are highly statistically significant. The mean marital distance among the primitive Bharias was 8.3 kms while it was higher (10.2 kms) among the non-primitive Bharias.

Differentials in Fertility and Family Planning

As given in Table 6.3, the fertility indicators, *i.e.* No. of Children born per couple, total fertility rate, mean age of reproductive life, mean age of first conception, mean age at first birth etc. do not differ considerably between the two groups of Bharias. The Infant Mortality Rate is higher in primitive Bharias (139.0) compared to the non-primitive Bharias (119.2). It may be noted that still birth rate is higher among the primitive Bharias. The fecundability at younger age is lower.

Table 6.3: Fertility and family planning behaviour

Sl. No.	*Indicators*	*Primitive Bharia*	*Non-primitive Bharia*
	Fertility		
1.	Mean no. of children born per eligible couple	2.7	2.5
2.	Mean ago of reproductive life (years)	28.3	27.4
3.	Mean age of first conception (years)	19.0	19.1
4.	Mean age of first birth (years)	20.5	20.0
5.	Abortion rate(%)	8.0	7.4
6.	Still birth (%)	2.0	0.4
7.	Total fertility rate	4.7	4.8
8.	Infant mortality rate	139.0	119.2
9.	Fecundability at age (18-19 years)	0.0434	0.0570
10.	Birth rate	33.3	33.4
11.	Death rate	14.9	13.3
	Family Planning		
1.	Eligible couples sterilized (%)	12.2	20.1
	Vasectomy	12.2	14.8
	Tubectomy	0.0	5.3
2.	Mean age of wife at the time of sterilization (years)	37.8	36.9
3.	Sterilization after mean no. of children surviving	2.9	3.2
4.	Sterilization after mean no. sons surviving	1.8	1.8

As regards to family planning behaviour, the percentage of sterilized couples is lower in the primitive Bharia tribe, while none was tubectomized among the primitive Bharia females, 5.3 per cent females adopted tubectomy among the non-primitive Bharias. There was no considerable variation in mean age of wives at the time of sterilization, number of children surviving to the sterilized couples and the number of sons surviving before adoption of sterilization.

Table 6.4: Comparison of Socio–cultural factors affecting maternal and child health and family planning practices

	Factors	*Primitive Bharias*	*Non-Primitive Bharias*
1	Concept of Pregnancy	Natural phenomenon	Natural phenomenon
2	Change in social behaviour among the pregnant women	(a) Do not see eclipse (b) Do not visit places of worship	(a) Do not see eclipse (b) Do not visit places of worship after 6 months of pregnancy
3	Utilization of PHC services	(a) They do not feel necessity (b) In case of any problem, they contact traditional Dai and Parihar	(a) Positive attitude towards antenatal care (b) In case of any problem, they visit to the doctor
4	Food items avoided during pregnancy	They avoid to take black chicken and mahua	No restriction
5	Preparation before delivery	The delivery is conducted in a separate room or outside the main house and it is cleaned before the delivery	No separate room
6	Delivery is conducted by	Usually elderly lady	Usually elderly lady
7	Delivery position	Usually in sitting and squatting position	Usually in sitting position
8	Cord is cut by	Sickle, old blade, axe	Sickle, new blade, old blade
9	Paste applied on stump	Broken Khapra or leaves of Chhatua	Broken Khapra
10	Colostrum	Not discarded	Usually discarded
11	Problems in lactation	(a) Papita chutni (b) Urad dal	Take medicines prescribed by the doctor
12	Awareness about family planning	Most of the people are not aware	Most of the people are aware

Differentials in Maternal and Child Health Care Practices

A comparison of maternal and child health care characteristics of the primitive and non-primitive Bharias is given in Table 6.4. As, evident from the table that the non-primitive Bharias are in advantageous position. They are more aware about the PHC services and they use these services more frequently when ever they need. Due to tribal transformation the non-primitive Bharias have adopted a number of practices which are not prevailing in the primitive Bharias, for example, use of medicines for lactation problems, consultation with a doctor in case of problems of delivery etc.

Thus, the primitive Bharias are socio-economically backward, the acceptance of family planning is lower and the practices related to maternal and child health care are more traditional than the non-primitive Bharias but the fertility performance of their woman does not differ considerably from the woman belonging to non-primitive group.

References

Abbad, A. (2000). Socio-cultural Determinants of maternal and child health care practices among the Bharias of Patalkot valley. Ph.D. dissertation, Rani Durgawati Vishwavidyalaya, Jabalpur.

Dixit, D.N. (1984). Plan for development of Bharias of Patalkot, Bulletin of Tribal Research and Developoment Institute. Bhopal.

Kurup (1986). Continuity and change in a little community, Concept Publishing Company, New Delhi.

Pandey, G.D. (1999). A study of effect of social factors on Human fertility and family planning with special reference to major primitive tribes of Madhya Pradesh. D. Litt dissertation. Rani Durgawati Vishwavidyalaya, Jabalpur.

Pandey, G.D. and Saxena, B.N. (1988). District Level Planning for Health and Family Welfare for Tribals. Demography India 17(2) pp. 216-226.

Tiwari, D.N. (1984). Primitive tribes of Madhya Pradesh - Strategy for Development. Govt. of India, Ministry of Home Affairs. Tribal Developments Divison, New Delhi.

7

Occupational Health: An Overview

P.K. Gangopadhyay
Regional Occupational Health Center (E),
Block -DP, Sector-V, Salt Lake City,
Calcutta (W.B.)

The environment, for practical purposes, may be divided into work environment and community environment. The workers are being exposed regularly to the different working environment like physical, chemical, biological, psychosocial, organizational etc. The community environment is also polluted by the different pollutants of the industries and or by the waste and other discharges of the community. The air and water, the minimum requirement for living are thus constantly polluted and the suffering of living organisms is on the increase. Crops are also badly affected . Due to mechanization in the industrial sectors and advent of newer things, the insult and assault to the environment is becoming so harsh that at times good number of lives are being lost. This is over and above the natural calamity. It affects the health status of both working and general population.

The occupational health deals the working environment and the health of the workers. It is the multidisciplinary approach to the recognition, diagnosis, treatment and prevention and control of diseases/injuries of occupational origin. It may be defined as: *it should aim at: the promotion and maintenance of the highest degree of physical, mental and social well being of all workers in all occupations; the prevention among workers of departures from health caused by their working conditions; the protection of workers in their employment from risks resulting from factors adverse to health; the placing and maintenance of the workers in an occupational environment adapted to his physiological and psychological ability and to summarise: the adaptation of work to man and each man to his job.* (Joint ILO/WHO Committee, 1950).

Health ⇔ Work

Occupational health is multidisciplinary in nature. Almost all the discipline of science are involved in it. The engineer, chemist, physicist, safety personnel, physician. occupational health specialist, epidemiologist, toxicologist, ergonomist, physiologist, psychologist, sociologist, statistician etc. are engaged here for evaluation, understanding and proper utilization of the discipline in the interest of the work ,workers and the nation. Considering the involvement of various disciplines a group approach is of immense value.

Occupational health is also a *shared responsibility*. In an organizational set up, particularly in an industry, various types of personnel are required for its functioning. Health hazards are inherent with the running of the industry unless proper precautionary measures are taken in appropriate manner in time. This is more so in our country where most of the work force are illiterate and are not aware of the hazards of work. Hence, discharging duties with tact, zeal and judgment by all responsible for the job is important in running the show.

It deals with working environment, the methods of working and conditions of work, mal-adjustment to work, mechanical risks, accidents, injuries etc. Considering the fact that most of the occupational diseases are *non curable*, a strict vigil by the responsible persons is very important. Some of the occupational diseases are *notifiable* and some are *compensable* as per the Acts promulgated by the Central Government and Rules framed by the State Governments from time to time. *Cost effective benefitness* and *health* of the working force are two vital sides needs special care in an industry. It is obvious that prevention of occurrence of disease is must in reducing/ controlling the suffering of the workers. The important of *"p"* in health is unparallel — *prevention* of occupational diseases leading to *preservation* and *protection* of the achieved health and ultimately *promotion* of the health status of the workers are the ultimate and prime objectives of occupational health practice.

Factors or variables responsible for *Occupational Health Hazards* in working community are many. Some of the important and commonly encountered factors are:

Physical	*Chemical*	*Biological*	*Ergonomic*
Noise	Dusts	Insects	Posture
Heat	Gases	Mites	Movement
Cold	Fumes	Fungi	Illumination
Vibration	Fibres etc.	Bacteria etc.	Visibility etc.
Radiation etc.			

The other important factors are *Organisational* and *Psychosocial*. They may act adversely or occasionally the action may be positive also.

Occupational diseases is said to be cause *specific*. It may also be affected/modified or even rarely caused by *non-occupational cause* like nutrition. Non-occupational pneumoconiosis in high altitude of Ladakh had also been reported (Norboo and Saiyed) in our country. So hazard specificity of occupational diseases may not be totally true. Moreover *multifactorial* causes are now advocated. There may be occupational health disorders super imposed with some infection or vice versa e.g. Silico tuberculosis.

Studies on work *promoting health* are very meagre. Studies should also focus on *work enrichment*. *Lack of work* may also be considered as hazardous to health. This may lead to psychosocial problems. *Mental health*, now a days, is very important in occupational health practice. Particularly, in our country, due to unrest among the workers, a stringent relation exists between the management and the labour force. The stress and strain on the executives leads to so many *work related diseases*. At times, emotional imbalance adds more to the agony. Poverty and other social problems lead to different types of *bad habits* to grow and the workers become victims of different diseases. In some cases they also loose their mental balance. Suicidal deaths are not very uncommon among the working group in our country. *Migration of labour* is another problem. The workers are staying away from their family and resorted to some unsocial activities and ultimately acquire diseases which may also be transmitted to their family members.

Target Organ

Occupational exposure or the insult and assault of the working environment on the specific organs of the body and its response to it result in pathogenesis of the disease. This is related with routes of entry of the substance in question and the affinity of the substance for the organ. It may be *inhalation, ingestion,* and through *intact skin.* Solubility of the substance helps its entry and settling property too helps to reach the specific organ. Of course, as stated the *dose* is very important for pathological process to set in. The specific effect as such depends on reaction of the substance with the concerned tissue, body's host or defence mechanism, the dose and duration of exposure, solubility of the substance, the cumulation effect, if any, etc. The organ(s) where the effect of the environmental exposure usually herald the onset of ill health is/are target organ(s). The effect may be on single organ or system or may be on multiple organs.

Differential Diagnosis

For proper diagnosis of occupational diseases the proper authenticated history of exposure for the life time, the environmental exposure, the other investigational reports are essential. The differential diagnosis to be kept in mind and investigations to be done accordingly. Clinical finding may also be helpful.

Pre-Placement Medical Examination

This is needed for screening and to have base line information of the workers concerned. As the proper and detailed history in respect of a worker is usually lacking, so it is very important. This can help in judging suitability of a worker for the specific job. Though it is not practised in our country, yet it has got immense value. It also helped in detecting ill-health. Sepecific investigations may be undertaken for specific group of jobs.

Periodic Medical Examination

In occupational health practice periodic medical examination is a must. This helps in assessing deviation of health from the previous health status. It may also help in judging the detectable early possible changes of health derangement and to formulate the action plan accordingly. In fact detection of earliest possible change when it is reversible or the damage is minimum is the prime objective of occupational health practice. In this case the subject may be shifted or alternate arrangement could be made for him. Preservation of the informations are essential for this purpose.

The Permissible Limits

Threshold Limit Value (TLV- TWA, STEL), MAC (Maximum Allowable Concentration) in respect of substrances used in the industry to be kept in mind. These are prescribed limit values for the pollutants/substances under which the worker may be permitted to work for the specific duration (40 hours a week) without any adverse health effect.

Occupational Diseases

The working environment may have adverse effect on the health of the worker. Depending on the type of agent, duration of exposure, quantum/concentration of exposure, the toxicity of the substance, the pathological processes set in the target organ(s) etc. the disease will express out involving single or multiple organ(s). The inhalation of dust may lead to occupational lung disorders like Pneumoconoisis; chronic exposure to lead may cause plumbism, exposure to high level of noise may be responsible for noise induced hearing loss etc.

Work Place Hazard Control

The main purpose of this is to minimize the exposure of the workers to the potential hazardous substances. This mainly consists of environmental control and personal protection. Control of working environmental conditions at source by using different engineering techniques - the best choice. If the measures are not sufficient to control then personal protection in additional to engineering control to be instituted. The effectiveness of the control measures need to be evaluated at certain interval of time and accordingly further necessary action may be taken to keep it under control.

Safety

It is well known that safety to be taken care at the first instance. Safety measures reduce load on medical side as well as number of accidents; make employees aware of the dangers, precautions taken, protect, preserve and promote health. This also helps to tackle any eventuality by them.

Rehabilitation and Resettlement

This is one of the important aspects where personal attention of the responsible persons is very much important. Disabled employees to be rehabilitated and or resettled involving different processes/ measures like: (a) Medical and or Surgical intervention, (b) Physiotherapy, (c) Re-location or re-placement in a suitable site or job, (d) Re-training, (e) Re-adjustment to family and society, (f) Counselling etc.

Health Education

It is needless to emphasize that our work force to be imparted with basic knowledge and education initially. Later on they will be taught regarding the effect of exposure on their health and what are the precautionary measures to be taken by him for protecting and preserving his health as well as of his fellow workers. This is essential for upliftment of working community. The awareness programme should also be extended to the management group. This acquired knowledge will help management to look into the safety and health aspects and to adopt the suitable remedial measures accordingly.

Legislation

It has already been mentioned that the Central Government has promulgated Factories Act and the State Government has framed Factories Rules for the safe running of the industrial units. The responsibilities of the owner have clearly been laid down in this act and rules. Similarly for mines the Mines Act is there. Notifiable and compensable diseases have also been mentioned in these documents. Guidelines have been given for availing of the necessary facilities for the benefit of the workers.

Recommendation

On the basis of the results, its interpretation, conclusion, implications, suitability of application, cost effective benefitness etc., to be viewed seriously while going for recommendation. Pre-employment and periodic medical examination, proper report keeping and its up-gradation, follow up of the workers as per requirements, periodic evaluation of environmental conditions and adoption of remedial measures (if any) are the important aspects needing special consideration. Emphasis needs to be laid on proper health education of the workers and is a must with special reference to adverse effects of dust and other environmental conditions.

However, the workers are not getting their due recognition and the benefit which are just for their sacrifice and devotion specially in terms of the working condition and health. Mr. Kofi A. Annan, Secretary General of United Nations said that though the "*United Nation's General Assembly recognises the right of all people to just and favourable conditions of work*", yet unfortunately hundreds of millions of people around the world are employed in conditions that deprive them of dignity and value. He further said that 250 million accidents occurred every year and out of this 330000 were fatal; 160 millions cases of Occupational diseases reported yearly. Ultimate approximate economic loss is about 4 per cent of world's gross national products. (Editorial, Asian Pacific news letter on occupational health and safety, vol. 4 (3), 1997).

Chemical production of the world grown by 400 folds from 1939 to 1997 and showing a growing trend. Presently 10 million chemicals are in the American Chemical Association's computer registry and 500 to 2000 chemicals are added each year (1988). But the safety measures and ill effects of most of these chemicals are unknown.

Occupational Disease Identification

For diagnosing occupational diseases like following schedule to be kept in mind:

A. Assessment of work environmental conditions.

B. Evaluation of the health status of the workers.

The comprehensiveness of the assessment is essential for arriving at a diagnosis.

a) Environmental Assessment

The different environmental conditions like dust, heat, humidity, air velocity, noise, illumination etc. to be measured using standard equipments following standard acceptable methods. This will give the magnitude of the problem of environmental conditions.

b) Evaluation of the Health Status of the Workers

This is to be undertaken as per the following programme:

1. Recording/collection of history with special emphasis on

detailed occupational history. Occupational history should include the details of each occupation with duration in each job. This is to be noted preferably in chronological order. For this purpose a standard proforma to be prepared and to be tested for its validity. The confidence of the subject to be gained and then the questionnaire to be administered for eliciting the information.

2. Depending on the assessment of environmental conditions, the history and the possible target organs(s) affected the different investigations to be undertaken, *e.g.*
   (i) Pulmonary Function Tests like Vital Capacity, Forced Vital Capacity, Forced Expiratory Volume, Forced Expiratory Volume in one second etc
   (ii) Haematological and Biochemical investigations
   (iii) Radiological examination
   (iv) Electrocardiography
   (v) Other sophisticated investigations depending on the type of expected outcome of the exposure
3. The differential diagnosis to be kept in mind and if required, other investigations may also be undertaken to exclude the other possibilities.

Occupational Lung Disorders

The effect of occupational exposure to the respiratory tract is well known. The exposure to the different types of pollutants may result in changes in the lung structure and function. It may be *reversible* or *irreversible, acute* or *chronic* depending on the *type of pollutants, dose* or *quantum of exposure, the duration of exposure, the toxicity of the substance, the body defence etc.* Individual *state of health, susceptibility, sensitivity* and *existing infirmity,* if any, may need to be taken into account. It is of importance to note that 'Silicosis' is the oldest most Occupational Disease known to the human civilization. Occupational Lung Disorders cover wide varieties of diseases with varying pathology. The specific disorders are mainly *Pneumoconiosis, Occupational Asthma* including *Byssinosis, Extrinsic Allergic Alveoiities, Non-neoplastic disorders, Cancers* etc. The causative agents may be *Dusts, Smoke, Fumes, other Physical and Chemical agents, Metals* etc.

The route of entry is usually by *inhalation*. The most important is the *air borne particles*. The inhaled particulates are *filtered/separated, captured* and *deposited* at different levels of the respiratory tract by *sedimentation, inertial impaction, interception* and *diffusion*. Of course the anatomical/structural configuration is also playing its own role.

Of the different occupational lung disorders Pneumoconiosis shared the major portion. It gives rise to specific changes in the lung parenchyma. It is a generic name and under its umbrella different diseases are there according to the causative agent responsible for the pathological changes. Zenker in 1866 introduced the terminology 'Pneumonokoniosis' for the lung disease caused by dust. The word simply means 'dusty lung'. Proust in 1874 modified the terminology to 'Pneumoconiosis'. There are varieties of definition of pneumoconiosis. Most of the definitions are based on compensation purpose. However, considering the permanent alteration/changes in the lung structure the Pneumoconiosis may be defined as *"the non-neoplastic reaction of the lungs to inhaled mineral or organic dust and the resultant alteration in their structure excluding asthma, bronchitis and emphysema"*. The pleura should not be included while diagnosing pneumoconiosis. Some of the examples are:

Dust Type	Reaction with lung	Disease
	Reticulin formation/mild	Siderosis-iron
	No fibrosis	Baritosis-Barium
		Coal Workers' Pneumoconiosis-early stage - coal dust
	Granuloma	Beryllium Diseases
Mineral	Collagenous nodular	Silicosis-free silica
	Fibrosis	Coal Workers' Pneumoconiosis-coal dust
		Mixed dust fibrosis
	Collagenous diffuse	Asbestosis - Asbestos fibres
	Fibrosis	Talc Pneumoconiosis
		Beryllium Disease
	No fibrosis	Bagassosis
	Acute extrinsic alveolitis	Farmers lung
	Interstitial pneumonia	Mushroon workers lung
Non-mineral organic		
	Collagenous fibrosis	Bagassosis
	Diffuse interstitial	Farmers lung
	Pulmonary fibrosis	Bird fanciers' lung

Coal Worker's Pneumoconiosis

The occupational health problems of the coal workers are well known. In general, the respiratory morbidity, skin problems, sexually transmitted diseases and helminthiasis are usually encountered. The specific health effect is due to inhalation of coal dust causing pneumoconiosis. Of course, it takes years to develop. The data available regarding the prevalence of Coal Workers Pneumoconiosis (CWP) in our country varies from 0.8 per cent to 18.8 per cent. But the studies were not comprehensive and reliability of the data is also questionable. This may probably be due to non-standardised techniques of diagnosis and faulty selection of samples. However, presently, the diagnosis of Coal Workers Pneumoconiosis has been standardized at the International level. The disease is notifiable and compensable.

Definition of CWP

Coal workers' pneumoconiosis may simply be defined as *"Occupational disease caused by prolonged retention in the lung of abnormal amounts of dusts derived in the first instance from coal mining operation". "Coal workers' pneumoconiosis (CWP) and carbon pneumoconiosis are specific diseases resulting from the inhalation and deposition in the lung of carbonaceous dust and the lung's reaction to the dust so deposited"* as defined by NIOSH, 1987. Simple coal workers pneumoconiosis and complicated pneumoconiosis or progressive massive fibrosis or masses pseudo-tumorales are the two forms.

Diagnostic Criteria

Simple coal worker's pneumoconiosis is without any symptoms and signs, if there are no predisposing diseases. Complicated pneumoconiosis may be having some symptoms and signs in the later stages. The diagnosis of CWP is based mainly on the X-ray chest with appropriate and authentic life time occupational history. The X-ray chest to be taken with a 300 mA X-ray machine in large films with appropriate kV (with or without grid) so as to get a clear cut picture of small pneumoconiotic shadows. Quality of film must be good. The X-ray taken to be compared with standard ILO films for classification of the X-rays by three independent readers. Pulmonary function test is of little value in some cases as the workers also may have other respiratory morbidities like chronic bronchitis and emphysema. Smoking is also one of the factors needing consideration. However, for differential diagnosis some common conditions mimicking simple CWP like tropical eosinphilia, military tuberculosis, sarcoidosis or histoplasmosis etc. to be kept in mind.

Collection of History with Special Reference to Occupational History

This is one of the most important aspects needs special attention. As our working community are mostly illiterate, special care must be taken to record authentic, reliable history. Standard proforma (preferably precoded) to be prepared, tested and validated prior to collection of information. Due stress to be given for noting occupational history in chronological order with duration of work in each job. This may be in the same organization or different organization and if required, proper description of the job to be noted for better understanding. Smoking history is very important. Quantum of smoke, duration of smoking and if left, duration of leaving to be recorded. Standard questionnaire may be included, if needed. Before collection of history, report to be established and the confidence of worker must be gained.

Respiratory Morbidity

It is documented that coal workers are victims of respiratory ailments. In addition to general medical history collection, standard respiratory questionnaire may be added to elicit the occurrence of chronic bronchitis. In a study the author observed that the main subjective complaints were dry cough (18.15 per cent), expectoration (16.88 per cent), breathlessness (28.44 per cent), chest pain (16.55 per cent) and haemoptysis (7.50 per cent). The above complaints are found to increase with age and duration of exposure. Chronic bronchitis was present to the extent of 8.83 per cent. Almost similar trend of increase of chronic bronchitis with age and work duration was noted.

Clinical Examination

Coal workers' pneumoconiosis is reported to be without any characteristic physical signs. Usually in simple CWP no significant sign in clinically detected. But at times, in few cases detectable clinical sign may be there. In complicated pneumoconiosis some signs may be detected, particularly at the later stages. Demonstrable clinical signs may be there because of non-pneumoconiotic persistent respiratory morbidities or in case pneumoconiosis is associated with other disease conditions. There are distinct possibilities that smoking may help in developing symptoms and signs. It is difficult to correlate/explain the cause of chronic bronchitis and dyspnoea. The irritant effect of coal dust may lead to hypersecretion of mucus in bronchitis but whether it is possible to produce obstruction of the airways sufficient to cause dyspnoea is difficult to explain.

Haematological Examination

For the diagnosis of CWP routine haematological tests is of no value. In case of other associated diseases or non-pneumoconiotic persistent respiratory morbidities, the tests may be of help. In our country in the mine areas tropical eosinophilia is commonly encountered with. It may also be helpful in differential diagnosis. Cases and in case of definite pneumoconiosis figures were chronic bronchitis - 2.36 per cent and dyspnoea 9.03 per cent.

Electrocardiogram

The ECG is an important investigation to know the prognosis of the penumoconiotic subjects. There may be right ventricular hypertrophy (RVH), with or without heart failure, particularly in case of complicated pneumoconiosis. Chronic or pulmonate or RVH is a late feature of pneumoconiosis. For routine purpose its use is limited in our country.

No specific treatment of coal worker's pneumoconiosis is available, but the complications, if any, to be treated accordingly.

Silicosis

The disease is most commonly found out and infact is taking sizeable number of lives in our country. Though the disease is notifiable and compensable, yet the diagnosis is not being made properly as such the figures are not depicting the real picture. It is caused by inhalation of free silica. It may be divided into four different types:

Pure or Nodular Silicosis

This is collagenous nodular fibrosis caused by inhalation of siliceous dust having proportionately high free silica content. Initially small discrete opacities seen and later or large nodular or irregular shadows developed. Egg shell calcification of hilar lymph nodes may be present.

Mixed Dust Fibrosis

It is ill-defined, irregular stellate fibrosis and is due to inhalation of siliceous dust containing mixture of free silica and inert mineral (*e.g.* iron oxide) with proportionately less free silica content. Discrete small nodular or irregular opacities similar to that of fibrocaseous tuberculosis.

Diatomite Pneumoconiosis

It is mainly diffuse interstitial pulmonary fibrosis caused by siliceous sedimentary rock diatomite containing diatomaceous silica. Initially linear or round opacities and later on circumscribed homogenous density shadows may be seen.

Acute Silicosis

It is alveolar silico-lipoproteinosis with diffuse interstitial pulmonary fibrosis and develops rapidly most probably due to siliceous dust with unusually high free silica content. Here mature silicotic nodules are either few or absent. There may be ground glass or small rounded opacities in the lower lung fields.

It is a fact that silicosis predisposes to the development of tuberculosis. In silicosis there may not be any symptom at the initial stage but advanced radiographic signs may be there. As the disease advances there may be cough, small quantity of sputum, breathlessness. Usually chest pain is absent. If there is complication other symptoms will be there accordingly. In silicosis cases detectable clinical signs are conspicuous by its absence unless otherwise complications are there and the signs of complications will be there. Lung function is of no help for diagnosing silicosis. Advanced cases may show impairment of lung function. Chest radiography is the only diagnostic tool in life. Standard procedures to be followed for taking radiograph of the chest. The radiograph to be interpreted comparing with the ILO standard films. Usually upper halves of the lungs are involved. Sputum may be examined for AFB. In acute silicosis cough, sputum, haemoptysis, chest pain and breathlessness may develop comparatively quickly. Clinical signs like clubbing, cyanosis, pleural rub, crepitation may be there. Restrictive impairment in lung function may be observed. Radiography has its own role of play. No specific therapy for silicosis is practised. However, complications may be treated accordingly.

Asbestosis

It is bilateral diffuse interstitial pulmonary fibrosis caused by fibrous asbestos mineral dusts. It is a fibrous mineral silicate. Usually it takes long time to develop the disease. Intially brethlessness on effort followed by even at rest, dry cough (late stage), small sputum, and occasional pain chest may be complained of. Persistent pain chest may be indicative of malignancy. The physical signs may be clubbing of fingers and toes, impaired expansion of chest wall. The *most important sign is persistent bilateral fine to medium basal crepitations in mid or early inspiration and continue to its end but not heard in the expiration.* Sputum for asbestos bodies is of no value. ESR may be high. Lung function tests may be of help along with history, signs and X-ray chest. Restrictive type of lung function impairment may be detected. It also help in assessing progression of the disease. Radiography is the most important investigation. For this purpose PA and suitably angled right and left anterior oblique views are desirable. Lower halves of the lung field is important for this. The radiographs may be interpreted as per ILO standard films by a panel of 3 interpreters. Irregular opacities are found but agreement of readers is difficult. Bilateral diffuse thickening of pleura may be there. Carcinoma of lung and pleural mesothelioma may also be developed. There is no treatment available at present for this disease.

Byssinosis

It is a gradation of respiratory symptoms due to exposure to the dust of cotton, flax, soft hemp, jute etc. Which range from acute dyspnoea, with cough and reversible breathlessness and chest tightness on one or more days of a working week of permanent respiratory disability due to irreversible airflow obstruction. It is a form of extrinsic asthma. Usually it occurs on the first day or Monday of the week and hence called Monday feeling. There are 4 clinical grades of Byssinosis. They are:

Grade C 1/2: Occasional tightness of the chest on the first day of the working week

Garde C 1: Tightness of the chest and/or difficulty in breathing on each first day only of the working week

Grade C 2: Tightness of the chest and/or difficulty in breathing on the first day and other days of the working week

Grade C 3: *Grade C 2* symptoms accompanied by evidence of permanent respiratory disability from reduced ventilatory capacity.

Smokers shows higher incidence of byssinosis. No physical sign usually present. Lung function is the most important investigation. Pre and post shift Pulmonary Function Tests and PFT for all the days of the week is very much informative. Progressive reduction of ventilatory capacity indicated by FEV_1 and maximum mid expiratory flow are sensitive tests. Radiography is of no help. Treatment consists of removal of worker and symptomatic treatment of symptoms.

Other occupational disorders like noise induced hearing loss due to exposure to noise; exposure to lead may cause lead toxicity; exposure to benzene bone marrow depression and leukaemia; exposure to azo-dyes cancer of urinary bladder, occupational dermatosis like oil achne, contact dermatits, leucoderma, melanoderma, cancer etc. due to exposure to irritants, sensitisers, metals.

In the new millenium the research on occupational health need to be shifted and following general points may be considered:

Shopfloor Management

(a) Population development and environmental development

(b) Impact assessment which will depend on affluency and technology

(c) Economics dealing with financial house keeping and Ecology dealing with environmental house keeping

(d) Sustainable development and sustainable use of energy/ demography/farming etc.

(e) Conciousness of environmental toxins like chronic low level pesticide exposure

(f) Deteriorating environmental conditions like deforestation

(g) Gene and environmental interaction; localisation of susceptible gene and the disease expression differs accordingly

(h) Transmission factors of various detoxicating genes for a particular organ

(i) Personal DNA card

(j) Individual tailoring of drug, suitability and unsuitability of environmental toxins

References

Asthama and Allergic Diseases: J Douglas Wilson, ADIS Health Science Press, Australia.

Cooper, J.K. and Jhonson, T.P. (1990). *Exercise capacity in coal workers' Pneumoconiosis: An analysis using causal modelling*, Br. J. Indust. Med.47.

Current Approaches to Occupational Medicine: Ward Gardner, J. Wright, Bristol.

Early Detection of Occupational Diseases: WHO, Geneva.

Environment and Health: Anthony, J. Rowland and Paul Cooper, Edward Arnold (Publishers) Ltd., London.

Environmental and Occupational Medicine: William, N. Rom, Little Brown and Co., Boston, Massachusetts.

Epidemiolgoy of Occupational Health: WHO, European Series No. 20.

Epidemiology in Medical Practice: Barker and Rose, Churchill Livingstone, Edinburgh.

Gardner, A.W. (1982). *Current approaches to Occupational Health 2*, John Wright and Sons Ltd., Bristal, Lnodon, Boston.

International Labour Organisation (1980). Guidelines for the use of ILO International classification of Radiographs of Pneumoconiosis, Occup. Saf. Health Series No. 22 (Rev. 1980), Geneva.

International Labour Organisation (1998). *Encyclopaedia of Occupational Safety and Health*, Geneva.

Krishnaswamy, S. (1972). *India's mineral resources*, IBH Pub. Co. Oxford, Delhi.

Lecture Notes on Occupational Medicine: Waldron, Blackwell Scientific Publication, Oxford.

Monitoring for Health Hazards at Work: Gill and Ashton, Grant McIntyre, London.

Morgan, W.K.C. and Lapp, N.L. (1976). *Respiratory disease in coal miners*, Am. Rev. Respir. Dis., 113.

National Institute of Occupational Health (Regional Centre, 1988-89). *Study of pneumoconiosis in underground coal miners in India*, Annual Report.

National Institute of Occupational Health (Regional Centre, 1991-92). *Study of pneumoconiosis in underground coal miners in India.*

National Institute of Occupational Safety and Health (1986). *Occupational respiratory diseases*, US Dept. of Health and Human Service, Washington D C (Revised), USA.

National Institute of Occupational Safety and Health (1987). *Occupational respiratory diseases*, US Dept. of Health and Human Service, Washington D C (Revised), USA.

Occupational Asthma: Claude Albee Frazier, Van Nostrand Reinhold Co., New York.

Occupational Diseases: A Guide to Their Recognition. National Institute of Occupational Safety and Health, NIOSH, Cincinnati.

Occupational Epidemiology: Monson, CRC Press, Boca Raton.

Occupational Health Practice: Schilling, Butterworths, London.

Occupational Health: Harrington and Gill, Blackwell Scientific Publication, Oxford.

Occupational Health: Recognising and Preventing Work Related Disease and Injury (2000). Ed. B.S. Levy and David H. Wegman, Lippincot Williams and Wilkings.

Occupational Lung Disease J. Bernard L. Gee, Churchil Livingstone Inc. New York.

Occupational Medicine: Zenz, Yearbook Publication, Chicago.

Occupational Safety Management and Engineering: Willie Hammer, Prentice Hall Inc., New Jersey.

Parkes, W.R. (1982). *Occupational lung disorders*, Butterworths, London, 2nd edition.

Recent Advances in Occupational Health: MacDonald, Churchill Livingstone, London.

Regional Occupational Health Centre (Eastern) (1994). *Study of Pneumoconiosis in underground Coal Miners in India*, Final report, ROHC(E), Calcutta, India.

The Diseases of Occupations: Donald Hunter, Hodder and Stoughton, London/Toronto.

World Health Organisation (1994). *Harmful exposrue to mineral dusts*, World Health Forum, 15(2).

8

Occupational Impact on Health of Biri Workers

A Case Study of Simdega Sub-Division

Umesh Chandra Nath Tiwari
Department of Geography,
Birsa College, Khunti,
Ranchi

In the Ranchi Plateau of Gumla District the term "Forest Dweller" means Tribe of the region.

Forest have shaped the habitat and economy of the Tribe and common people of Simdega Sub-Division of Gumla District. The tribal people are as much as living part of the forest as the wild animals. Forest are the very fountain of life for the tribe. Most of the activities pertaining to food, housing, occupation and health are performed in the forest.

Upon forest resources depend a large number of economic activities like skilled and unskilled work. Some of the unskilled industries like Biri making are also based on forest resources. Forest are a source for livelihood in the form of edible and non-edible leaves plucking.

In recent years, new measures have encouraged to show practically new Geographical Research. Now a days, Geographical study of Environmental Conditions and Health Care is an addition in the field of Medical inter-related components:

(a) Occupation

(b) Nutritional status

(c) Health conditions

(d) Health care

Natural and Cultural environment in which nature of work is one of the main (Biological Processes) controlling factors for the health conditions of workers. Therefore, the environment in which people live has much to do with the types of diseases they suffer from Environmental impact is the most important aspects of the pollution problems caused by rapid increasing of industrialization. Mankind suffer from polluted environment that's result ill health. Our respiratory system is directly affected by air pollution when we inhale polluted air enter into our blood and affects our blood streams, lungs, heart and nervous system causing many chronic diseases. Social and cultural conditions not only influenes the process of learning but also helps in determining individual health characteristics, which are mainly responsible for man's total environment.

Biri making is one of the major house hold forest based industries of Simdega Sub-Division, Gumla District, which provides employment to tribal and poors. Biri making influenced the workers by process, materials used and quality of waste disposed to a great extent and also to the environment.

The majority of poor classes of the study area have been forced to choose biri making as a main base for their livelihood. The lower income groups of muslim and Tribal families are engaged in this scattered industry all over the region.

Area of Study

Simdega Sub-Division of Gumla District (23°30′ - 23°40′N and 84°40′ - 84°50′E) are situated in the southern part of the district. Physiographically, study area is the part of Ranchi Plateau or rejuvenated peneplain. Simdega Sub-Division is partly different from the rest of the peneplain which is a slightly lower level (1500-2000 ft.). The area composed of Archaen system of granite-gneiss rock. The lowest temperature of winter is 8°-9°C in January and the summer mean temperature varies from 23°C to 37°C in the Simdega Sub-Division. The forested south western tract in Simdega generally gets over 64 cm to 100 cm rainfall during the last rainy season from the Bay of Bengal branch of the monsoon.

Physiographic and climatic differences of the area have given birth to moist deciduous type of forests. Some common species of the deciduous forest are Semal, Sal, Asan, Tendu, Karanj, Gamhar and Mango etc. The study area depend primarily on road transport for both external and internal relations.

Population of the Simdega Sub-Division is composed essentially, very high share of Tribal population.

Objectives and Hypotheses

The main objectives of this study is to examine the co-relation the environmental conditions created by the Biri Making Industry and its affect on the health of the Biri Workers.

Keeping the objectives in mind the hypotheses have been formulated on the basis of the socio-economic, demographic and occupational characteristic are examined.

These hypotheses are:

(a) There is close relationship between Environment, health and occupation.

(b) This occupation of Tribal is totally forest based.

(c) Diseases caused by this occupation is mainly due to Tobacco consumption habit.

(d) Working efficiency decreases to lack of proper food availability.

(e) Due to lack of job opportunities they are forced to continue their work.

Methodology

The present study is based on the primary and secondary data. Field survey was conducted by the authors. The survey is based on the study of a number of families (35) from the different localities with various economic and religious background. To confirm the effects of working conditions and environment, 4 Government Doctors and 5 Private Doctors from the some localities were also interviewed to have their view along with the detailed information regarding the physical disorders of the workers. On the basis of personal observation I have detected the environmental condition of the workers of Biri Industries as follows:

(a) Large family size of the workers
(b) Illiteracy
(c) Poverty
(d) Customs and religious concepts which affects their living, food habits and health
(e) Lack of civic sense and amenities
(f) Physical negligence
(g) Poor dietary composition
(h) Unhealthy environment

In the study area, Biri workers come from different social economic and religious background. The customs and religious concepts has influenced the size of a family and crowd in the factory premises.

Table 8.1 shows Tribal and non-muslim common people have at least five-six members in each family but muslim workers have large family size than tribal and other workers. Generally, they have eight-nine members in the family.

Table 8.1: Community wise population of biri workers

Community	*No. of family*	*Population (total)*	*Male*	*Female*
Muslim	12	108	62	46
Tribal	14	77	49	28
Others	09	38	24	14
	35	223	135	88

Landuse

Physiography of the area is influence land utilization and occupational structure of the society. The land use pattern in the area shows that forest occupies about 63 per cent and about 28 per cent of the total geographical area is cultivated. Agriculture in this area is mostly rainfed which is spread over the period from June to September.

Monocropping pattern of the cultivation is due to non-availability of irrigation. Thus, the forest is the source of livelihood, source of occupation and other activities. People of the area is collected and used the minor forest produce as a occupation. Rice and millets are the main food crops.

In the study area, land is the back bone's for the people's livelihood. But the large number of families have no land and if some of them have a few plots that is not sufficient for their livelihood. After agricultural work workers have no work, ultimately they are forced to choose the forest based work.

Table 8.2 reveal that tribal women are also engaged in Biri making and support their families.

Table 8.2: Number of workers engaged in biri making age and community wise

Age group	*Muslim*		*Tribal*		*Others*	
	Male	*Female*	*Male*	*Female*	*Male*	*Female*
Below 15	08	02	07	-	02	-
15-30	16	07	22	06	08	03
30-45	12	03	03	02	04	02
Above 45	06	-	01	-	01	-
	42	12	33	08	15	05

Literacy

Illiteracy and lack of knowledge is also responsible for the economic condition, food habits, job opportunities and health care. Majority is illiterate. Some family are only non-matric. One thing is important that the females who are in charge of food are mostly illiterate. Literacy rate among females are below 10 per cent and males are approximately 22 per cent.

The majorities of families are not able for healthy food consumption due to large family size.

Food Habits

Workers food habits are controlled by the culture, income, food availability, agriculture production, literacy and family size.

Usually they take Rice, Gruel and Dal (Pulses) three times a day. A part from this muslims and common people children take Roti irregularly and tribal children take (marua) 'Mamaroti' when they feel hungry. Seasonal vegetables are used by all community but seasonal leaves are normally used by tribal. While animal food are occasionally used by muslims. Milk is completely absent in the routine diet in all community. Usually the workers take local made liquor like "Haria" and Jharuni (Mahua liquor) daily.

Thus, the diet is deficient in respect of many vital nutrients, especially for female workers and children. All these circumstances of Biri workers cause Mal nutrition. Table 8.3 shows community wise nutrient intake of Biri workers by conducting the diet survey and was converted into various nutrients on the basis of I.C.M.R. Bulletin.

Table 8.3: Nutrient intake of Biri workers (1999-2000 per head/per day)

Community	*Calories*	*Protein (Gm)*	*Calcium (Mg)*	*Iron (Mg)*	*Vita. A (IU)*	*Ribo. (B1)*	*Thia (B2)*
Muslim	1812	58.6	388	12.2	416	1.7	1.2
Tribal	1793	54.4	372	19.9	635	1.5	1.1
Others	1962	60.2	468	12.7	833	1.6	1.3

Source: Diet Survey.

On the basis of disease survey among the Biri workers it is clear that the respiratory system is a leading disorder. Total reported cases in the study area followed by Bronchitis, gastritis, chronic dysentery, deficiency disorders, anaemia are common in females.

In most of the cases the persons suffereing from disorder of respiratory system may have all the symptoms present in group i.e. cough, expectoration, pulmonary, haemorrhage, chest pain, breathing troubles, pneumonia, bronchitis, tonsils, abcess of lung, influenza, malaria and common colds etc. are the main reported troubles by persons suffering from respiratory disorders.

These respiratory disorders are mainly due to crowded atmosphere, polluted drinking water, poor sanitary conditions, open and poorly maintained drains, poor ventilation, poor and insufficient diet which add to the respiratory and infectious diseases.

The general housing condition of the Biri workers has been found not only insufficient but are also quite unhealthy. Most of the houses do not have proper light and ventilation. The occupants are never in the position to utilize the valuable vitamin D produced by the skin due to the action of Sun Rays. The economic conditions of Biri workers found very poor. Due to poor economic condition they are unable to take balanced diet and medicine regularly to improve their health condition. Some workers are suffering from acute bacillary dysentery can produce gastro entesities.

Teeth and gum diseases are very common among the Biri workers due to regular consumption of Tobacco and carelessness to their teeth. Females and children are suffering from anaemia due to poor diet. Multiple deliveries, religious rigidness and their primitive health care system creates fatal situation. Lack of personal cleanliness and surroundings provides conditions from skin diseases.

Generally food consumed by Biri workers do not provide the required nutrition on the basis of their working conditions and working hours, it is essential to maintain the capacity. Normally workers of Biri Industry are unable to take milk.

During the survey it has been also observed that cooking methods are unhygienic, defective and in most of the cases food is prepared near in a place of open drain or in a dark corner of the house which results in ill health conditions.

Conclusion

On the basis of whole analysis it can be concluded that Biri workers are insufficient to meet their working condition due to which the following deficiencies have been found common in them, *i.e.*

(a) Low energy food intake

(b) Presence in food insufficient amount of protein

(c) Low proportion of minerals

(d) Lack of variation in diet

(e) Deficiency of Vitamins.

The co-relation of the general distribution of diseases with the environment where Biri making is done dietary habits and social background are of great importance in the present context. Besides these customs and traditions and female's illiteracy also causes health hazards. Thus, we can say that natural and cultural environment are directly or indirectly influences most of the disorders.

Thus, we feel that to improve the working conditions of Biri workers it is necessary to implement all the schemes formulated by the Government with sincerity regarding:

(a) Child development.

(b) Nutritional programmes.

(c) Establishment Primary Health centre.

(d) Awareness about personal hygiene and sanitation

(e) Availability of proper housing.

(f) Industry must be far from residential areas.

(g) Education facilities for their children.

(h) Small family having one or two child must be encouraged to improve their living condition.

(i) Awareness about personal and community cleanliness.

(j) Proper drainage system and drinking water facility.

References

Ahmed Enayat (1965). Bihar (R.U.) p. 15.

Agarwal, S.N. *India's Population Problems,* pp. 1-18.

Ankroyed Gopalan Balsubramanium (1966). I.C.M.R. Special Report Series No. 42, NIM, Hyderabad.

Maj, J.M. (1950). *Medical Geography,* Geographical Review Vol. XL, New York.

Wadia,.D.N. *Geology of India*, p. 107.

9

Societal Development in Mining Areas

Naresh C. Saxena
Dean (P & RG),
Indian School of Mines,
Dhanbad - 826 004

Abstract

A study of the socio-economic profiles of seven coal mining complexes indicates that the societal development and the status of the society is far from satisfactory. In fact the complexes have different classes of society and lack basic facilities for a major percentage of their population.

Based on the Quality-of-life criteria developed by the author and others a methodology has been suggested for adequate societal development planning of mining complexes, which can as well be used for other industrial areas.

Preamble

Mining of minerals and coal is generally considered as an environmentally infriendly activity having severe impacts on the society in and around the mining areas. This becomes glaring if one looks into the societal aspects of the mining areas. But, is it an universal fact? The answer cannot be 'YES', as in many places in Europe the post mining societal scenario is better than the pre-mining scenario. If the European countries have achieved it why countries like India cannot achieve similar results is a matter of debate to develop a suitable system/methodology.

Outlined later is a methodology which in the author's opinion can be adapted in the Indian mining areas for the overall societal development.

General Characteristics of Mining Complexes

Mining of minerals is a site specific activity as the minerals are site specific resources. In India most of the mineral and coal resources are in forest-cum-agricultural areas and the general characteristics of these areas are as outlined below:

(a) Mining and associated activities are generally in remote rural areas.

(b) Mining has brought about distinct changes in the life styles of the local people.

(c) Many of the local people live at subsistence level, are poorly educated and are from traditional cultures.

(d) Invariably semi-skilled and skilled manpower is not available in ethnic people. Hence, for most of the mining activities manpower is brought from outside.

(e) Mining has made marked changes in land-use, ecology, water regime, etc. in the areas.

(f) Mining has increased the level of economic activities manifolds.

(g) Mining is only an intermediate use of land as after the extraction of minerals the land and, may be, the infrastructure is of practically no use to the mining companies.

Societal Impacts of Mining

Mining of minerals and associated activities have been causing various impacts on the society and the most important is the changes in the population dynamics which is depicted in Fig. 9.1. The various impacts that the activities causes are as listed here - under:

(a) Population changes

(b) Societal complextion changes and disintegration of families

(c) Cultural changes

(d) Decrease in sex-ratio

(e) Displacement of people

(f) Loss of livelihood

(g) Development of civic and infrastructural facilities

(h) Development of educational infrastructure

(i) Increased cost of living

(j) Better transportation, communicational and medical facilities

(k) Recreational facilities

(l) Changes in addictions

(m) Increased aspirations

(n) Increased income and also higher income disparity

(o) Employment potential and changes in the nature of employment.

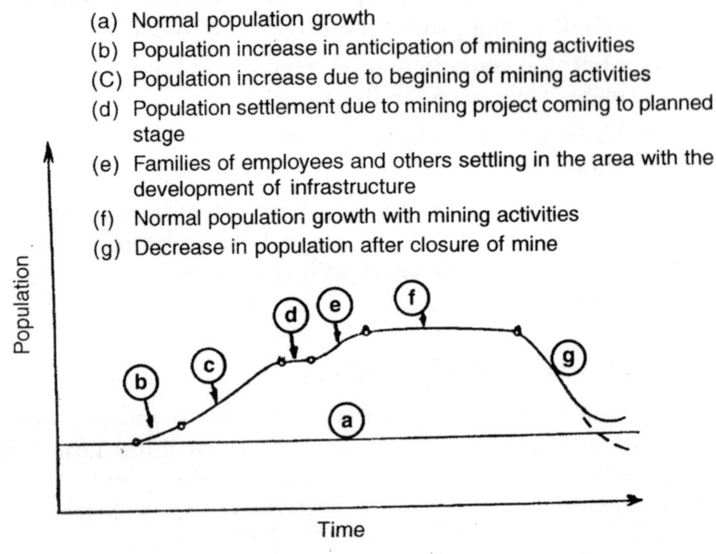

Fig. 9.1: Population dynamics in a mining area

The impacts are both advantageous as well as disadvantageous and with proper planning the disadvantageous impacts can be converted into advantages.

Socio-Economic Profiles of a Few Mining Complexes

The Centre of Mining Environment, ISM, Dhanbad conducted socio-economic studies in seven coal mining complexes in recent years. The salient features are outlined hereunder:

(a) There was a marked change in the structure of the society in all the complexes with dominance of the local language and nuclear families.

(b) More than 50 per cent of the families in all the complexes had more than five members with three and more children.

(c) The sex ratio varied greatly from 790 to 1004.

(d) The age-wise population pattern was as given below:

< 5 years	5.4 - 8.7 per cent
5 - 20 years	24.5 - 45.0 per cent
20 - 40 years	28.4 - 38.5 per cent
40 - 60 years	13.0 - 23.7 per cent
< 60 years	3.0 - 7.9 per cent

Thus, the population of capable workers (20 - 60 years) was about 44.0 to 62.2 per cent.

(e) The overall literacy varied from 56.8 to 72.0 per cent.

(f) In the five coal mining complexes in Andhra Pradesh 58.4 to 76.2 per cent people were from the same district. But in the two complexes in Bihar this percentage was 38.5 to 42.5.

(g) Housing status, water availability and sanitation facilities were far form satisfactory.

(h) Majority of the people were not satisfied with the medical facilities.

(i) Except in two complexes (both in AP) more than half of the families were spending more than 50 per cent of their income on food items.

The Quality-of-Life (QoL))of the families and the seven complexes was as given in the Table 9.1, which was assessed by the methodology developed by the author and others at CME, ISM, Dhanbad for mining and also, may be, for other industrial complexes (Saxena, *et al.*, 1999).

Table 9.1: Quality-of-Life (QoL) of the mining complexes

Sl.No.	QoL family	Complexes						
		1	2	3	4	5	6	7
1.	Poor (%)	60.0	54.0	7.7	16.0	8.8	13.4	38.6
2.	Fair (%)	35.0	37.0	82.0	61.0	74.2	80.4	55.7
3.	Good (%)	5.0	8.0	9.3	21.0	8.2	5.0	5.7
4.	Very Good (%)	—	1.0	1.0	2.0	8.8	1.2	—
5.	QoL Complexes	Poor	Poor	Fair	Poor	Fair	Poor	Poor

(a) Bhowra area of BCCL in JCF

(b) West Bokaro mining complex of TISCO in West Bokaro Coalfields

(c) Kothagudem area of SECL

(d) Ramagundam area of SECL

(e) Manuguru mining complex of SECL

(f) Mandamarry area of SECL

(g) Bhopalpally mining complex of SECL

It is noticed that out of the seven coal mining complexes only two had an overall fair quality-of-life. This indicates that there is an urgent need to look into the societal development of the mining complexes and initiate necessary activities. The criteria for fair quality of life of the families and the complexes is outlined in Tables 9.2 and 9.3

Table 9.2: QoL ratings of the families and complexes

Rating of families		Rating of complexes	
QoL score	Rating	Per cent of families Having poor QoL score	Rating
< 6	Poor	> 10	Poor
6 - 20	Fair	5 - 10	Fair
20 - 40	Good	< 5	Good
> 40	Very good		

Table 9.3: Desirable features of fair QoL (Unit: A family of five members)

Sl.No.	*Feature/parameter*	*Desirable status*
1.	Housing	Two bedrooms, one living room, kitchen, varandah, and toilet in an area of about 60 sq. m
2.	Water	Water availability of 200-200 l/day conforming to IS 10500
3.	Food and nutrition	Calorie intake of 2200 (minimum) per head per day for adults
4.	Health and safety	Not more than 50 mandays lost due to illness etc.
5.	Sanitation	Sanitation facilities having septic tanks with flushing
6.	Education	100 per cent literacy. A minimum education upto matriculation for all eligible
7.	Medical	Access to hospital facilities to all
8.	Transport and communication	Good roads with bus service and public telephone booths
9.	Fuel and energy	Availability of coal as fuel for all in coal mining complexes. Preferable LPG
10.	Own transportation	Ownership of a two wheeler (scooter or motorcycle)
11.	Income	Minimum take home salary/income of Rs. 6000/- per month per family
12.	Recreation	TV and cinema
13.	Family assets	Total family assets of > Rs. 2,50,000

Methodology for Societal Development

Before discussing the methodology for societal development being suggested by the author it would be appropriate to look into the various measures being undertaken by the mining companies, which are briefly outlined here under:

(a) The industry is implementing "resettlement and rehabilitation" (R&R) packages for the "project affected people" (PAPs).

(b) Villages have been adapted for their over all development.

(c) Employment opportunities have been provided to the local people not only in mining but also in other directly and indirectly associated activities.

(d) Medical, educational and infrastructural facilities have been developed.

(e) Efforts are being made to preserve the historical, cultural and religious heritages of the areas, etc.

A closer look on the societal structure of the mining complexes reveals that the colonies developed by the mining companies have distinctly better QoL compared to surrounding villages and the R&R locations.

In order to achieve the goal of societal development in mining complexes necessary emphasis is required to be given to this at the planning stage itself. The various steps that may be involved are briefly outlined below:

Step 1

Development of accurate socio-economic profile of the complexes and assessment of the QoL prevailing and the aspirations of the society. This should cover both the core and buffer zones.

Step 2

Identify the Project Affected People (PAPs) on the basis of the minimum possible land requirement for mining and associated activities. For this purpose it is essential to study the existing land-use and design the surface layout so as to cause minimum possible damages to the land.

Step 3

From amongst the population of the area identify the capable workers who could be gainfully employed in the mining and associated activities.

Note: Close interaction with the local people is essential for obtaining their confidence and hence getting the desired information from them. Sometimes it may be necessary to raise the aspiration of the people.

Step 4

Assess the impacts of mining and associated activities on society and land.

Step 5

Devise rehabilitation and resettlement (R&R) Schemes, indigenous people development programmes (IPDP) and community development programmes (CDP) keeping in view the desired QoL at various stages. In this planning the following need special attention:

(a) As far as possible in R&R schemes there should not be any provision of jobs in less of land and loss of livelihood. R&R policy and hence the schemes should be based on the principle of bringing back the PAPs at their original place after due reclamation and making compensation for the loss of livelihood in the intervening period.

(b) Post mining land-use plan and mine closure should be planned to facilitate the implementation of R&R schemes and also to make the area suitable for living and development.

(c) All actions required for implementation of IPDP and CDP may not be the responsibility of the mining companies. Other State/Central Government departments and local organizations should be encouraged to do the jobs concerned with them.

(d) Local people should not be treated as second class citizens of the areas and with their confidence already available due to interaction with them right from the beginning their involvement in development the R&R, IPDP and CDP would help.

(e) The mining companies should establish their rapport by demonstrating the good intentions.

(f) The post mining land-use plan should also be developed in consultation with the local people.

It may be noted here that the developoment of post mining land-use plan and hence proper mine-closure plan are essential ingredients of a good societal development plan. The mine-closure plan may spread over the period from the time of

(a) starting of backfilling in opencast mines, and

(b) management of subsided areas in underground mines.

Step 6

Design mining activities with the minimum possible land requirement and also minimum possible societal costs of environmental pollution, *i.e.*, integrate environmental management measures in mining and associated activities/operations so that the air pollution, water pollution and land damage potential of mining is the least.

Step 7

Design infrastructure facilities in such a manner that the ethnic people and rehabilitated and resettled people do not feel neglected. This would help in general improvement of their outlook and may be QoL.

During designing the mining activities with various other activities outlined above care should be taken that the plans are practically implementable.

The societal development planning should be need based and should address to the needs of the complexes. This can be expected to yield better overall results and post mining development of the areas. The assessment of the needs of the complexes and their population is thus an important aspect for development planning. In this connection the following need special mention.

1. Plan the surface layout such as to affect –
   (a) minimum possible displacement of people,
   (b) minimum possible loss of livelihood,
   (c) least disturbances in land use, etc.
2. In the socio-economic profile look for –
   (a) the persons, unskilled, semi-skilled and skilled, who could be employed in the project activities,
   (b) the persons having various skills, who could become resources to industry by way of doing associated minor activities, *e.g.*, carpentry, blacksmithy, weaving, etc.,
   (c) the persons who would continue to use the land and other resources to produce goods and services, etc.
3. Develop the education system and facilities –
   (a) for meeting local and regional needs,

   (b) for developing understanding amongst the people regarding the local and regional resources so as to practically utilize them for overall development, etc.

4. Develop the local resources for –
   (a) primary activities' enhancement,
   (b) attracting economic investment,
   (c) tourist interest, etc.
5. Develop civic facilities and infrastructure for –
   (a) the overall development and to meet the requirement of the entire population,
   (b) achieving the desired QoL of the complexes, etc.
6. Expose ethinic population to the desired developmental scenario so as to gain their confidence and to get access to their special needs, if any.

Some of the mining complexes may have some exclusive needs and for the 'need based development planning' (NBDP) it is necessary to assess these needs, *e.g.*, –

(a) in the remote areas development of transport network, *i.e.*, roads and railway may help to a great extent,

(b) in a predominantly agricultural areas development of scientific agricultural methods and agro-based industry can be expected to help,

(c) water and effluent management in almost all mining situations would be useful,

(d) in forest areas minimization of forest land's use for mining and development of forest based industries could be planned, etc.

Conclusion

Community development planning with QoL considerations and also with the involvement of the ethinic people seems to be an important aspect of overall societal development in mining complexes.

The QoL based concept can be effectively used and in some special situations it may become necessary to make modifications in the concept.

In the mining complexes overall societal development should not be sole responsibilities of the mining companies. Their role should be limited to taking care of PAPs and to providing some support to others nearby. The overall infrastructure and civic facility development for the total community should be done by respective State and Central Government departments and local autonomous organizations.

Large scale socio-economic data should be generated in respect of the mining complexes so that the various developmental strategies can be fine-tuned wherever necessary.

Acknowledgement

The author is thankful to the Director, ISM for according permission to present this article. The views expressed in the article are those of the author.

References

Kumar, P. (1996). *Socio-Economic Profile of West Bokaro Mining Complex of Tisco.* M.Tech Dissertation, ISM, Dhanbad (Unpublished).

Kumar, P.S. (1998). *Socio-Economic Profile of a Mining Complex of Singareni Collieries Co. Ltd.* M.Tech. Dissertation, ISM, Dhanbad (Unpublished).

Pratap, A. (1988). *Socio-Economic Profile of a Mining Complex - A Study of Mandamarry Area.* M.Tech Dissertation, ISM, Dhanbad (Unpublished).

Prusty, B.K. (1996). *An Investigation into the Socio-Economic Profile of Bhowra Area of Jharia Coalfield.* M.Tech Dissertation, ISM, Dhanbad (Unpublished).

Rao, G.S. (1998). *Investigation into socio-economic profile of Manuguri Mining Complex.* M.Tech Dissertation, ISM, Dhanbad (Unpublished).

Rao, P.V. (1997). *Assessment of socio-economic profile of Kothagudem Mining Complex.* M. Tech. Dissertation, ISM, Dhanbad (Unpublished).

Rao, Y.S. (1997). *An Investigation into the Societal Profile and Noise Status in Ramagundam Mining Complex.* M.Tech Dissertation, ISM, Dhanbad (Unpublished).

Saxena, N.C. (1994). Impacts of Industrialisation on the Tribals in Jharkhand, *Social Change*, v. 24, n. 1&2, pp. 119-123.

Saxena, N.C., Pal, A.K., Kumar, P. and Prusy, B.K. (1997). Quality of living in two coal mining complexes in India. *Proc. 2nd World Mining Environment Congress,* Katowice, Poland.

Saxena, N.C., Singh, G. and Pal, A.K. (1999). *Social Environment of Coal Mining Complex.* A Monograph by ENVIS Centre on Mining Environment. CME, ISM, Dhanbad (Unpublished).

10

Coal Mine Spoils Reclamation with Legumes and Grasses: An Environmental Field Study

M.P. Singh and Namita Sinha
Department of Forest Sciences,
Birsa Agricultural University,
Kanke, Ranchi - 6

Abstract

An experimental field study of using legumes (*stylosanthes humilis*) and grass (*Pennisetum pedicallatum*) in ratio of 1:2 was carried out on coal mine spoils at Kuju Coal Fields. The study was initiated on July, 1996 and the pH, conductivity, soil fraction, organic carbon (OC), total nitrogen (total-N), available nitrogen (av-N), Phosphorous and potassium were monitored on yearly basis up to July, 1998. At the end of three years soil fraction increased to 45 per cent, pH improved from 7.8 to 6.8. The reduction of conductivity from 0.73 to 0.30 mmhos/cm was noted. The increasement of OC was 14 per cent and 79 per cent and at the end of second and third year respectively. The concentration of av-N was increased to 562 kg/ha after three years. No significant accumulation of available phosphorus was observed. Exchangeable-K concentration was increased to 127 ppm after three years.

Introduction

A raw substrate of coal mine spoil is not a soil but only a matrix of mineral materials, of very poor physical structure, no organic matter and probably very restricted amount of plant nutrients. If a vegetation cover can be established, it will begin the process of soil formation. Mine spoil always lacks in nutrients particularly nitrogen and phosphorus. The use of legumes and grass mixer for the initial establishment of vegetation cover mine spoil has been practised by many coal companies abroad (Holechek, *et al.*, 1982, Nicholas, *et al.*, 1982).

As commonly known, legumes are members of plant family leguminosae and ranges from small herbs (*Desmodium*), to shrubs (*Trifolium, Medicago*) to trees (*Leucaenea, Acacia etc.*) and fix atmospheric nitrogen by bacteria (*Rhizobium*) nodulated on the leguminous root. Different legumes need different species of bacteria to produce maximum amounts of usable nitrogen (Lyle, 1987). Legumes grown on mine spoil help to increase soil nitrogen as mine spoil is low in organic matter and nitrogen. The rate of nitrogen fixation by different legumes is as follows - Alfalfa (*Medicago sativa*) - 224 kg/ha; Red clover - 129 kg/ha, Kudzu - 123 kg/ha, Soybeen - 112 kg/ha, Cowpea - 101 kg/ha etc. (Donahue, 1990).

Thus, legumes are grown to aid the soil forming process, to add nitrogen to the soil and to provide a bulky palatable component to the sward. The legumes greatly improve fertility by accumulating nitrogen especially where manuring is not practised (Richardson, 1977).

The present work has aimed to study the: (i) usefulness of grass *Pennisetum pedicellatum* and legume (*stylosanthes humilis*) mixer as initial colonizer in coal mine spoil, (ii) study the rate of nitrogen fixation by leguminous plant on spoil.

Material and Methods

Test Species

The legumes 'Stylo' (*Stylosanthes humilis*: Leguminousae-Papilonaceae) was collected from Birsa Agricultural University (BAU) Ranchi in the form of seeds. The *Rhizobium* culture was inoculated at culture room of the University. The Legume is perennial, semi-shrubs types with trifoliate leaves and attains 0.6 to 0.9 m height root: shoot ratio is 6:1 in initial stage of growth.

The grass species used is commonly knwon as 'Dinanath grass' (*Pennisetum pedicallatum*: Poaceae was collected from Birsa Agricultural University, Ranchi in form of seed. The grass attains a height of approx. One meter, is an annual, drying out in the month of November-December. The root is fibrous and restricted to a maximum of 15 m spread on ground and the seeds are feather type.

Experimental Plots

Five plots (3m × 2m) were prepared by levelling the coal spoil at the onset of monsoon (June 1996). Three plots were revegetated with legume and grass mixture (1:2) and two plots were left for natural colonisation. No fertilizer application or watering was done throughout the experiment, but small quantities of top spoil (1-2 cm depth) was applied at the start of experiment. By July 1998, the main plot as well as the periphery of the plots (for 0.6 to 0.9 m distance) was fully covered by legumes only.

At the end of three years those plots left for natural colonisation were dominated by *Hyptis* sp. (Leguminosae). *Tridex procumbens* (Asteraceae) and *Cleome viscosa* (Capparidaceae). But in revegetated plots even after three years, no natural colonising species have invaded.

Spoil Sampling

Spoil was sampled using a split tube coring tool operated manually. In July, 1996 seven points were randomly located and spoil core to a depth of 20 cm was extracted. In July, 1998 duplicate core samples were collected adjacent to the previous sampling points. On each occasion samples were air dried, mixed thoroughly and about half kg was collected after quartering.

Spoil Analysis

The spoil pH and Electrical conductivity (EC) were determined in a suspension of 1:2-5 (spoil water; w/v) in pH meter and conductivity meter respectively.

The organic carbon was determined by Walkley and Black method (Potassium dichromate reduction and titration with ferrous ammonium sulfate) and total nitrogen by micro kjeldhal method. Available nitrogen was estimated by alkaline permaganate, method. Available phosphorus from spoil was extracted by Bray's reagent (spoil - to extractant ratio of 1:5) and determined by molebdenum blue method. Exchangeable potassium was extracted by 1 (N) ammonium Acetate solution spoil to extractant ratio of 1:5 and was determined by Flame photometer. Water extract was prepared by 1:5 ratio (spoil water).

Result and Discussion

The study (Table 10.1) indicates that after three years of

revegetation with legumes, quality of mine spoil was improved as discussed below:

Table 10.1: Physico-chemical properties of spoil material

Sl.No.	Parameters	Legumnous plots (0-10 cm) Initial	After 2nd year	After 3rd year	Control old spoil Plot (0-10 cm)	Dumps (0-10 cm)
1.	Colour	Light black	black	black	black	blackish gray
2.	pH	NA	7.78 (7.6-7.9)	6.86 (6.7-7.0)	6.6 (6.5-6.8)	6.3 (4.8-6.3)
3.	Conductivity (mmhos/cm)	NA	0.73 (0.70-0.75)	0.30 (0.28-0.33)	0.16 (0.15-0.16)	0.33 (0.15-0.43)
4.	Soil particles (%)	27	32.2 (28.5-34.2)	48.0 (45.0-52.0)	46.5 (43-54)	44.5 (41-51)
5.	Organic carbon (%)	0.43 (0.38-0.50)	1.05 (0.90-1.15)	1.90 (1.6-2.0)	1.0 (0.63-1.4)	0.8 (0.4-1.7)
6.	Ignition loss (%)	NA	5.72 (4.0-6.5)	9.88 (9.6-11.0)	NA	NA
7.	Total Nitrogen					
	(a) in (%)	0.008 (0.005-0.01)	0.0173 (0.015-0.020)	0.164 (0.16-0.17)	NA	NA
	(b) in PPM	80	173	1640	-	-
	(c) kg/ha	180	393	3708	-	-
8.	Available - N					
	(a) in %	0.0025 (BDL-0.0025)	0.007 (0.006-0.008)	0.025 (0.027-0.032)	0.017	0.006
	(b) in PPM	25	70	225	175.2	66.7
	(c) kg/hs	56	157	562.5	395	150
9.	Available - P					
	(PPM)	Traces	1.6 (1.25-2.0)	4.75 (4.0-5.5)	1.8 (0.7-2.8)	0.9 (BDL-1.6)
10.	Potassium (PPM)					
	(a) Soluble	NA	11.2 (9.0-15.5)	32.66 (28.0-38.0)	15 (10.2-18)	12.5 (10.2-15)
	(b) Exchange-able	NA	11.2 (55.5-60.0)	32.66 (120-140)	15 (105-140)	12.5 (63-71)

NA = Not analysed, BDL = Below detection level.

(a) *Soil Fraction:* Particle less than 2 mm in size is termed as soil. Initial soil fraction in the soil was 32 per cent and the end of three years it was 47 per cent, which is an increase of 45 per cent. In control plots and old spoil dumps, the amount of soil fraction were found to be 40 per cent and 44 per cent respectively at 0-10 cm depth. The presence of even as little as 20 per cent fine material has dramatic effects upon vegetation establishment and growth (Down, 1974). This increament is largely due to reduction of run off from spoil surface. If may be noted that to reduction of run off from spoil surface. It may be noted that *Penisetum* grass has very extensive fibrous root system which anchor the loose spoil material. The legumes (*Stylosanthes, humilis*) tested was perennial type and has a very good cover (greater than 100 per cent)

(b) Improvement of pH after 3 years of establishment with grass and legume mixtures, pH of the surface spoil (0-10 cm), changed from 7.8 to 6.8. The pH of 5.0 or higher is most suitable for nitrogen fixation (Lyle, 1980). Bacteria will not penetrate many of the Legumes if the spoil pH is below 5.0.

(c) *Improvement of Conductivity:* The change in conductivity from 0.73 mmhos/cm to 0.30 mmhos/cm was observed after three years and in conductivity range was found in between 0.2-0.8 mmhos/cm which is optimum for plant growth (Beer, 1964).

(d) *Organic carbon (OC):* Initially OC level was very low (0.44 per cent) with passage of time the level of OC was increased and reached to 1.9 per cent at the end of three years. The increment of OC was 14 per cent and 79 per cent for second and third year respectively. In the control plot, after the end of three years, OC level were found 1.0 per cent. Thus, the enhancement of OC level by grass and legumes mixtures was by a factor of 1.9. The OC acculation in natural dumps at 0.10 cm depth was found to be an average of 0.81 per cent, with a variability of 76 per cent.

This high variability of OC can be attributed to the heterogeneous nature of spoil. Initially rapid increment of

OC level was due to accumulation/decomposition of legumes and grasses on the surface. As organic matter increases with time nutrients retention capacity also tends to increase (Caspall, 1975).

(e) *Mineralisable/Available Nitrogen:* It has been well established that (Dancer, *et al.*, 1977, Marrs *et al.*, 1981) nitrogen accumulation and building up of nitrogen cycle is the most crucial factor in soil development in mined out areas. In present study, initially Av-N was very low (56 kg/ha). After two years accumulation Av-N was measured to be 157 kg/ha and it increased to 562 kg/ha at the end of three years. When accumulation of Av-N by legumes was compared with that in the control plot and old dumps, the enhancement factors were found to be 1.42 and 4.0 respectively in 0-10 cm depth. It has been reported that most legumes can fix 50-150 kg of N/year *Coppin* and *Bradshaw* (1980). Li *et al.* (1988) have studied the rate of nitrogen accumulation in Southwest Virginia mine spoils by leguminous plant and found the same to be 60 kg/ha and 16 kg/ha at 0-5 and 5-10 cm depth respectively. They observed that below 10 cm depth there was very little accumulation of nitrogen. The main source of nitrogen accumulation was due to symbiotic nitrogen fixation. The interrelation between nitrogen level, organic carbon and time period are significantly correlated. A minimum level of 700 kg/ha of spoil nitrogen is essenital for self sustaining plant communities. To achieve the figure of 700 kg/ha natural colonisation takes about 50 years to reach the level, whilst legumes will take 5-10 years only (Coppin and Bradhshaw, 1980).

(f) *C:N Ratio:* The C:N ratio in soil have been reported in the range of 11.5:1 (Sing and Datta, 1988) 8.41 (Rokima and Prasad, 1989) 11.5:1 (Jacksonm, 1973). The C:N ratio in the spoils was observed to 60:1 and 11.6:1 after second and third years respectively. The improvement in C:N ratio is due to higher accumulation of organic matter by mineralisation process, (Donahue, *et al.*, 1990). According to Lannings and Williams (1979), only when C:N ratio is 25:1 or less there will be a significant net release of nitrogen from microbial activity.

(g) *Available Phosphorous:* The main source of available -P in mine spoil is the breakdown of spoil materials and precipitation from atmosphere (Coppin and Bradshaw, 1982). The concentration of av-P in mine spoils, after two years was very low (1.25-2.0 ppm). The low level of P may be attributed to the low release of P from shale and fixation of a considerable portion of P shale (Gould, 1979). Many workers observed that (Vamamto, 1975; Safaya, 1979) total P of mine spoil may equal or exceed those of the unmined soils, but the plant av-P in almost all the spoils as invariably in the deficiency range. At the end of three years, av-P has been increased to 4.7 ppm. The P concentration below 20 ppm indicates that fertilizer is needed. Value of 5-10 ppm shows that a moderately high rate of phosphate should be added and 0-5 ppm indicated that a very high application rate is needed (Gammell, 1990).

(h) *Available Potassium:* The concentration of available. K, after three years was found to be 127 ppm which is sufficient for plant growth. According to Gammell (1990) Av-K concentration of 100 ppm is sufficient for plant growth, 50-100 ppm indicates moderate deficiency range and 0-50 ppm indicates high deficiency. The accumulation of Av-K in control plot and natural dumps were 114 ppm and 66.5 pm at 10 cm depth respectively.

Conclusion

After three years of experiments with legumes and grasses on coal mine spoils the following conclusion could be drawn:

(a) Organic carbon is increased by factor 1.9 at the end of three years.

(b) Available - K was found to be sufficient after three years.

(c) The optimum C:N ratio in mine spoils could be achieved after three years.

(d) Level of available - P even after three years was found in the deficient range.

Thus, for establishment of long-term reclamation progress on mine spoils following recommendations could be suggested:

(a) To improve the C:N ratio, use of legumes is the cheapest method to increase nitrogen level.

(b) To ameliorate the harsh environmental condition of high temperature and low field moisture on mine spoil surface soil amendment such as saw dust, hay etc. should be used.

(c) During revegetation study Av-N should be monitored seasonally and once it reached a level of 700 av-N/ha tree plantation could be started.

(d) Addition of phosphatic fertilizer is essential for mine spoils, reclamation becuase even after three years av-P level is in the deficient range.

(e) Initial spreading of small quantities of top soil, say, of 2 cm depth on soil could signifcantly ameliorated the spoil medium.

(g) Lastly, becasue of highly variable nature of spoil material the size of the experimental plots have to be increased for better scientific analysis of the results.

Acknowledgement

The author is greatful to Ministry of Environment and Forests for providing the funds for this study. The authors would also like to thank CCL authorities and staff for providing the site and necessary facilities during the experiment.

References

Beer, E.F. (1964). Chemistry of soil. Oxford and IBH Pub. Co. Limited, Calcutta, p. 501.

Copping, N.J. and A.D. Bradshaw (1982). Quarry Reclamation. Mining Journals Books London, p. 18.

Caspall, F.C. (1975). Soil development on surface mine spoils in western Illinion in 3rd Sym. on Surface Mining and Reclamation, Vol. 2, National Coal Association, Lousiville, Kentucky, p. 221-8.

Down, C.G. (1974). The relationship between colliery waste particles size and plant growth. Environ. Conser. 1(4), 281-4.

Donahaue, R.L., R.W. Miller and J. Shickluna (1990). Soils an introduction in kailin mining wastes in Cornwell. I, an introdution to soils and plant growth. Prentice Hall of India Ltd., New Delhi, p. 130.

Dancer, W.S., Handley, J.F. and Bradshaw, A.D. (1977). Nitrogen accumulation in Laolin mining wastes in Cornwell. I. Natural Communities, plant and soil (48), 153 168.

Holechek, J.L., E.J. Depuit, J. G. Coenenberg and R. Valdez (1982). Legume establishment of strip mined lands in southern Montena., J. Range Manage., 35(3).

Jackson, M.L. (1973). Soil Chemical Analysis, Prentice Hall of India Pvt. Ltd., New Delhi, p. 183.

Lyle, E.S. (Jr.) (1987). Surface mining reclamation manual Elsevier, N.Y. p. 19.

Marrs, R.H., Roberts, R.D. and Bradshaw, A.D. (1981). Ecosystem development in reclaimed China clay wastes, I. Assessment of vegetation and recapture of nutrients. J. of Applied Ecology (17), 709 717.

Rokima, J. and Prasad (1989). Kinetics of Nitrogen mineralization in calcareous soil as influenced by integrated nutrient supply system.

Sing, O.P. and B. Dutta (1988). Organic carbon and Nitrogen status of some solis of Mizoram occuring at different altitudes.

11

Assessment of Dust from Opencast Iron Mine of Kiriburu (Jharkhand)

Abhijit Dutta, Anjali Dutta, Sunita Dutta and P.N. Pandey
Environment Laboratory,
Department of Zoology,
Ranchi University,
Ranchi

Abstract

Dust is produced due to mining operation and anciliary activities. The data collected indicates that dust continues to be a major air pollutant in the mining area of Kiriburu. Inspite of limited water sprinkling operation there is detrimental impact of dust on the health of mining personnel. Observations also indicate that a maximum of 576 $\mu g/m^3$ is prevalent in the affected area which may cause eye and lung diseases among mine workers.

Introduction

Quality of air of Kiriburu Iron Mine area is singnificantly affected by some of the human activities such as burning of fuels, mining operations, vehicular movements etc. Introduction of dust, smoke and noxious gases produced as a result of human activities modify the air quality to a large extent; (Derek, 1987).

There are two properties of a mineral that should be taken into account to generate dust brittleness and hardness. Brittleness measures the inherent tendency to form dust while stress required to cause deformation representes hardness (Stern, 19771; Das, 1977;

Down and Stock, 1978). However, no dust formation will take place unless a force is applied; (Dutta, 1999).

Mining activities at Kiriburu, located in the western limb of Bonai range in Saranda Reserve forest with prolific Sal and Teak trees, have brought about significant change in the environmental quality. The excavation and deforestation have made the soil brittle. Added to this there are various other mining activities, one significant is the production of huge quanitities of overburden, that are potential source of enormous amount of dust having adverse impact on the health of the mine workers (Kumar, 1982; Sinha, 1984; Owen, 1985; Rao, 1989).

The present article tries to assess the amount of dust generated at various parts of the Kiriburu area and the consequent environmental effect.

Materials and Methods

Assessment of Suspended Particulate Matter (SPM)

SPM was measured using high volume air sampler, of make Envirotech High Volume Sampler, Model No. APM 411, Serial No. 393 DT 192 with 101-104 m^3/min of air passed through weighed GF/A. Whatman filter paper for an average of 7 to 9 hrs. run and the change in weight of the filter paper was measured to calculate the SPM in $\mu g/m^3$ using the following methodology of Harrison and Perry (1986):

For SPM measurement the technique of "time integrated sampling' is most popular, where samples are collected for a period of time and the concentration of SPM is calculated by dividing the total quantity of dust collected by total volume of air.

Calculation

Air flow rate (Q) in m^3/min = Va/T

Where Va = air volume at atmospheric pressure in m^3

T = time of flow in minutes.

$$\text{Volume of air sample (V) in } m^3 = \frac{(Q_1 + Q_2) \times T}{2}$$

Where, Q_1 = initial air flow rate in m^3/min

Q_2 = final air flow rate in m^3/min
T = sampling time

Mass concentration of SPM

$$\text{Mass concentration (C) in } \mu g/m^3 = \frac{EF - WI \times 10^6}{V}$$

where, EF = final mass of filter paper in gram
WI = initial mass of filter paper in gram
V = air volume sampled in m^3, correct at 25°C.

Sample Location

Samples test were carried out in different parts of the mine and adjacent areas. Five sample sites were chosen near sources of dust pollution. These are - Haul road, Petrol pump, Hopper, Field office and Rest house area.

Observation

The table mentioned shows the SPM concentration at Kiriburu Mine over a period of 3 months from March, 1998 to June, 1998. It was observed that there were many factors responsible for dust generation in the mining areas. The table indicates that:

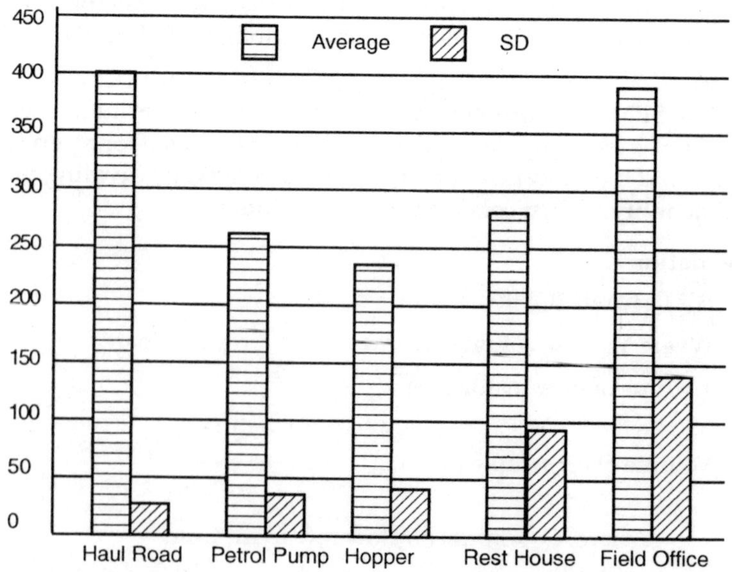

Fig. 11.1: Average value of SPM and SD

Table 11.1: SPM levels at various location of Kiriburu iron mine

Location	*Date*	*Total run hrs. (Hour & Min.)*	*SPM in µg/c.cm*
Haul Raod	11.3.98	7.9	395.00
	22.4.98	8.15	436.67
	13.6.98	8.16	356.92
	17.6.98	7.23	425.00
	18.6.98	7.6	399.68
Average		7.8	402.65
SD (Standard Deviation)		0.38	30.90
Petrol Pump	12.3.98	7.9	307.00
	23.4.98	8.00	266.00
	14.6.98	7.8	197.32
	19.6.98	7.58	250.85
	22.6.98	6.9	274.91
Average		7.6	259.22
SD (Standard Deviation)		0.38	35.99
Hopper	13.3.98	7.07	241.00
	24.4.98	7.6	167.68
Average		7.35	204.34
SD (Standard Deviation)		0.28	36.66
Rest House	15.3.98	6.7	171.00
	25.3.98	8.03	421.10
	16.6.98	8.20	273.87
	21.6.98	7.6	237.24
Average		7.6	275.00
SD (Standard Deviation)		0.58	91.63
Field Office	24.3.98	8.0	201.00
	24.3.98	7.08	382.00
	15.6.98	7.23	469.36
	20.6.98	7.7	576.99
	23.6.98	8.34	321.57
Average		7.6	390.18
SD (Standard Deviation)		0.53	142.89

The average value of SPM in Haul-road, Petrol pump, Hopper, Field office and Rest house areas were 402.65 μg/m^3, 259.22 μg/m^3, 204.34 μg/m^3, 390.18 μg/m^3 and 275.00 μg/m^3 with standard deviation of 30.9, 35.99, 36.66, 142.89 and 91.63.

The data indicate that the dust level is highest in the Haul road area while the large standard deviation of 142.89 in the Field office indicate huge fluctuation of dust level in different months. It was found to be 576.99 during June and 201.00 during March.

The sampling sites where mettled surface roads are present, like Petrol pump area and Hopper. SPM levels were found to be low 259.22 μg/m^3 and 204.34 μg/m^3. The low SPM levels in these area was also because they are away from the active mining sites and indicates settling action of rain and vegetation.

It was further observed that dust storms are rare in these area because it is in the midst of Saranda forest zone and the wind velocities are low. The past wind data indicates that wind speed above 22 km. are rare and accounts for only one day/year on an average. Dust storms accounts for 2 days/year *vide* Chaibasa Meteorological record.

Discussion

The data gathered indicate that dust is the principle air pollutant in the mining area of Kiriburu. Observations of the effectiveness of the limited water sprinkling activites reveals that inspite of 200-230 kl of water sprinkled per day, dust levels at various sites are quite high.

There are several reasons of dust generation in the study area. These are blasting at the mine faced followed by lifting of ore by power driven shovels; loading of ore by the shovels into the dumper trucks; dumping of ore into the gyratory crushers of the beneficiation plant; movement of the dumper trucks and other vehicles on the unsurfaced access roads; drilling operations; operation at the primary crusher; dumping of overburden at different sites; land clearance; overburden removal, bench development etc.

The environmental nuisance of dust depends also upon air flow of a velocity sufficient to carry it from the point of origin. According to Elson, (1987) a wind speed of more than 16 ft/sec dust of about 100 km will be raised from stationary dry surface and carried away downward for 600 ft, wind speed of 300 ft/sec can transport dust 2600 ft. Thus, on internal mining roads (Haul road) dust is often a continual potential for environmental nuisance.

The detrimental impact of dust, especially on health of mine workers, may lead to eye and lung diseases and also to reduced visibility as reported by Kumar (1982); Sinha (1984); Owen (1985) and Rao (1989).

References

Das, D.C. (1977). Soil conservation practices and erosion control in India, Soil Bulletin of FAO, 33: 11-50.

Derek Elson (1987). Atmospheric Pollution, Basil Blanck well td. pp. 220.

Down, C.G. and J. Stock (1978). Environ. Impacts of Mining. Allied Pub. pp. 371.

Dutta, Anjali (1999). Ecological studies of Mining Activities - A case study of IISCO Iron Mine. Unpub. Thesis, pp. 79-86.

Dutta, A. And Dutta, S. (1999). Dust pollution from opencast Iron Mine and its Impact - Trans Zoo. Soc. East. Idnia 3(2): 1999.

Kumar, V.K. (1982). Kanpur city: A study in environmental pollution. Tara Book Agency, Varanasi.

Nair, P.K. and Sinha, J.K. (1987). "Dust and Coal Workers" Pnenmoconiosis and some control measures, Survivar IIT. KGP pp. 16-20.

Owen, O.S. (1985). Natural Resource Conservation: An ecological approach, Macuillau, N.Y.

Perry Harrison (1986). Handbook of Air Pollution Analysis, Chapman Hall.

Rao, M.N (1989). "Air Pollution", Tata McGraw Hill, pp. 221-254.

Sinha, S.P, (1986). Urban environ. and contemporary ecology. The Indian Pub. Pp. 179-184.

Stern, A.C. (1977). "Air Pollution" Vol.-II, A.P. pp. 215-613.

12

Dust Sources and Control Techniques in Longwall Panels

Shambhu Sharan
Mining Engineer,
CMPDI, Kanke Road,
Ranchi - 834 008

Abstract

At present, longwall system is the most productive, cost-effective, and safest system of underground mining under favourable conditions. As a result of major strides made in the advancement of longwall technology during the past two decades, the production and productivity of longwalls have increased several folds worldover. With the steadily increasing rate of output from the longwall panels, the problem of airborne respirable dust has also increased presenting difficulty in complying with the prevailing statutory standard. Such difficulty arises primarily because the developments made in the area of controlling dust have been overshadowed by those made for enhancing the production potential of longwalls. To combat this problem, an integrative approach towards controlling dust is necessary. Such an approach requires identification of dust sources and assessment of their contribution in contaminating the face ventilating air. Shearer is considered the major source of dust at the longwall face whereas the stageloader - cursher as the major source outbye the face. This article presents the study conducted on some longwalls in the US to determine the dust generation rate of the shearer and contribution of dust sources outbye the longwall face in contaminating the face intake air. It is observed that although the dust generation rate of the shearer varied within a wide range, most of the times it was below 200 mg/tonne of coal mined. Among, the outbye sources, stageloader - crusher was the dominant source contributing upto 66 per cent of the dust level of the face intake air. Also, a review of the means of controlling dust generated by these two sources is presented in this article.

Introduction

Longwall mining has demonstrated its superiority over other underground coal mining methods under favourable conditions in terms of production, productivity, resource recovery, and safety. As per the Coal Age's International Longwall Census, the number of fully mechanised longwalls in major coal producing countries around the world exceeds 1700. Being the most productive, cost-effective and safest form of underground mining, longwall system will be the dominant underground mining system in the foreseeable future. Though this technology has been adopted in India also, but it has to go a long way in exploiting the potential of this technology to the maximum extent.

Due to the conceptual leaps and advances made in longwall mining technology, the production and productivity of longwall faces have increased several folds in the countries pioneer in longwall mining. According to US Mine Safety and Health Administration (MSHA) data, the average production of longwall sections increased from approximately 2400 tonnes per shift (tps) in 1988 to more than 3600 tps in 1995. About 25 per cent of the longwalls were capable of producing consistently over 6000 tps (Jankowski and Organiscak, 1996). During 1998-99, only 62 longwall faces US produced over 200 million tonnes of coal. The Twenty mile Mine in Colorado (Western US) during 1998-99 produced 8.5 million tonnes of clean coal from a single longwall face. The mine holds the current monthly record of producing 850,000 tonnes of clean coal from a single longwall face, and also the 10-hour shift production record of 26,548 clean tonnes of coal by making 32 complete passes on 256 m wide and 2.6 m high longwall face. As much as coal flow of 6000 tph has been recorded over the belt conveyor. Another mine in the western US, West Elk Mine, has achieved several excellent shift, daily, and monthly production figures on the 290 m wide longwall face in 3.3 m - 3.96 m thick coal seam (Paschedag and Martin, 1999).

Such performances of longwalls are associated with generation of large quantity of dust by different sources being released into the ventilating air current. With increasing production and productivity, the respirable dust concentration in the mine air is also increasing. A study by Jankowski and Organiscak (1996) indicates that an increase in shift production rates increases the dust levels at the shearer operator (Figure 12.1). Also, the fact that while 40 per cent of the operating longwalls in the US were in continuous compliance with the statutory dust standard of 2 mg/m^3 in 1990, only 28 per cent of the same in continuous compliance in 1995 is an indicator of serious dust problem in longwall operation. This is primarily because the strides made in enhancing the production potential of longwalls have overshadowed those made in the control of airborne respirable dust. While the mining industry is envisioning higher and higher rate of output longwalls, the dust control problem assumes even greater importance. In view of the dust problem, the control of dust from all sources at the working face and outbye the working face becomes critical. It is thus important to identify the sources of dust and estimate their contribution in contaminating the face ventilating air.

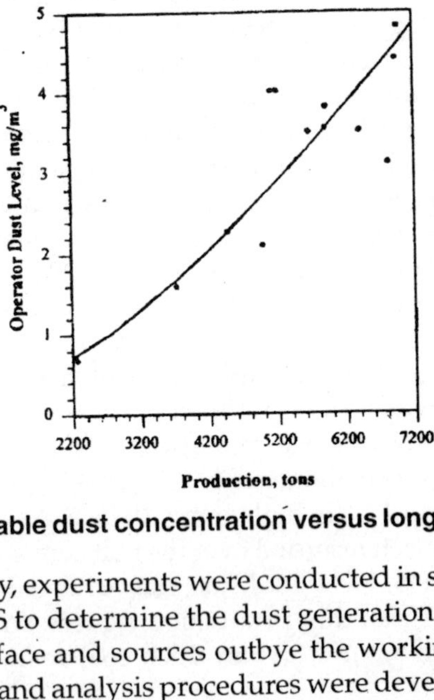

Fig 12.1: Respirable dust concentration versus longwall production

In this study, experiments were conducted in several longwall panels in the US to determine the dust generation rates of shearer at the working face and sources outbye the working face. Specific sampling plans and analysis procedures were developed to achieve this objective. Also, some of the dust control techniques to reduce the dust exposure of the face workers have been discussed.

Sources of Respirable Dust

Sources of airborne respirable dust in longwall panels are categorised as - sources at the working face and sources outbye the working face.

Sources at the Longwall Face

Shearer, support movement, and transport of coal along the flight conveyor are the sources of dust at a longwall face. Past studies by US Bureau of Mines (Jankowski and Organiscak, 1983; Jankowski and Hake, 1989; Jankowski, Jayaraman, and Organiscak, 1990) indicate the shearer while cutting is the major source of dust at the longwall face contributing from 40 per cent to 60 per cent (1.2 mg/m^3 to 9.8 mg/m^3) of the total respirable dust exposure of the shearer operator. In a number of field evaluations, instantaneous respirable dust level of 20-30 mg/m^3 were measured at the shearer operator's location during the headgate cut-out operation. Such high dust levels can have substantial effect on shearer operator's total exposure on high production longwalls where large number of cutting passes are expected. The shield movement contributes upto 33 per cent and transport of coal along the flight conveyor contributes upto 9 per cent of the total respirable dust exposure of the shearer operator.

Outbye Sources

Stageloader-crusher, panel belt, and panel-to-main belt transfer point are the outbye sources of respirable dust in longwall panels. Among the outbye sources of dust, the stageloader-crusher has been found to be the dominant source. At several longwalls surveyed by the US Bureau of Mines, more than 50 per cent of the shearer operator's dust exposure came from the stageloader-crusher (Jayaraman, Jankowski, and Organiscak, 1992).

A study by Organiscak, Jankowski, and Kelly (1986) in a two-entry longwall in which about 25 per cent of the intake air was brought up the belt entry, the respirable dust concentration in the belt entry upwind of the cross-cut where it joins the main intake was 2.4 mg/m^3 and its contribution in contaminating the face intake air was 0.5 mg/m^3. A study conducted by Haney (1996) in six longwalls in which the belt air was being used for face ventilation, the dust concentration in the belt entry approached or exceeded 1.0 mg/m^3 contributing in the range of 0.1-0.3 mg/m^3 in contaminating the face air. The contribution of primary intake ranged from 0.1-0.4 mg/m^3 and that of the stageloader-crusher was high ranging from 0.5-1.3 mg/m^3. The combined contribution of the outbye sources ranged from 0.6-1.7 mg/m^3. Thus, the contribution of outbye sources individually may not appear to be significant, but taken collectively the contribution of all outbye sources in contaminating the face intake air is certainly a matter of concern.

Details of Experiments

Experimental studies were conducted in six different mines. These mines were operating in different geological and mining conditions with different operational factors. Salient features of the longwall panels under study are summarised in the Table 12.1.

Table 12.1: Salient features of the longwall panels under study

Mine	*Panel*	*Average Depth* (m)	*Average cutting* height (m)	*Face width* h (m)	*Web Width (mm)*	*Drum dia.* (mm)	*No. of entries*	*No. of Experi-ments*
Mine 1	Panel-1	90	1.5	288	813	1422	4	3
	Panel-2	90	1.5	270	813	1422	4	6
Mine 2	Panel-1	120	1.7	190	914	1575	3	2
	Panel-2	120	1.7	279	990	1575	3	6
Mine 3	Panel-1	180	1.7	300	1016	1422	3	3
	Panel-2	180	1.7	300	1016	1422	3	4
Mine 4	Panel-1	180	1.8	262	914	NA	3	3
Mine 5	Panel-1	457	1.8	256	813	NA	4	1
Mine 6	Panel-1	457	1.8	280	813	NA	4	2

In Mine 1, Mine 2 and Mine 3, belt entry air was not used for face ventilation, whereas in the other three mines belt entry air was used for face ventilation. The experiments conducted at Mine 1, Mine 2 and Mine 3 were primarily aiméd at determining the dust generation rates of the shearer, and those conducted in rest of the mines were aimed at determining the contribution of various outbye dust sources in contaminating the face intake air.

The experiments involved dust surveys, air quantity surveys and time-studies on shearer operation.

Sampling Plans

For Determining Dust Generation Rate of Shearer

In Mine 1, Mine 2, and Mine 3, five to eight fixed airbone respirable dust (ARD) sampling stations were established along the longwall face (Figure 12.2). The sampling package at each station consisted of a gravimetric personal dust sampler with nylon cyclone preseparator and a real-time aerosol monitor (RAM).

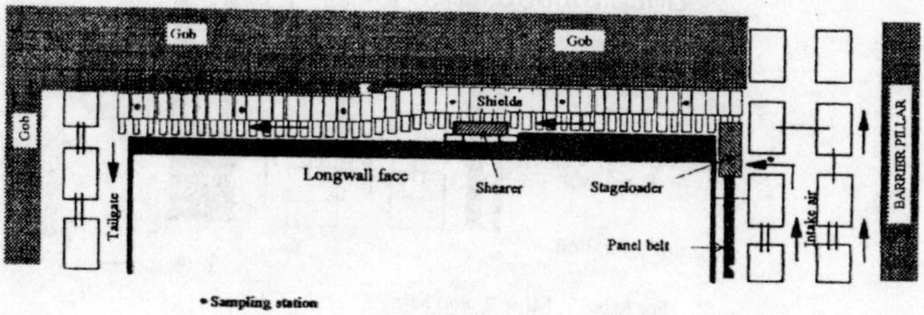

Fig. 12.2: Sampling plan at longwall face

For Outbye Sources

In Mine 1, Mine 2, and Mine 3, one sampling station was located in the intake cross-cut (station 1) and the second (station 2) was located at the face between 5th and 10th shield as shown in Figure 12.3. Gravimetric dust samplers were used at the sampling locations. In Mine 4, Mine 5, and Mine 6, in addition to the above two sampling stations, a third sampling station was established upwind of the stageloader-crusher in the belt entry (Figure 12.3). In these mines, in addition to the gravimetric dust samplers, RAM units were also installed at stations 2 and 3.

In the panels of Mine 4, Mine 5, and Mine 6, sampling stations were also established at regular intervals along the panel belt from head to tailpiece. At each station, gravimetric dust samplers were used.

To estimate the contamination by the panel-to-main belt transfer point, gravimetric dust samplers were located at two sampling stations - one in the main belt entry outbye the transfer point, and the other in the panel belt entry inbye the transfer point.

Data Analysis

Dust Generation Rate of Transfer

The instantaneous ARD concentration data recorded by RAM units installed along the longwall face, air quantity, and the shearer time study data have been used to determine the dust generation rate (R mg/tonne) of the shearer. Two different approaches have been used for the determination.

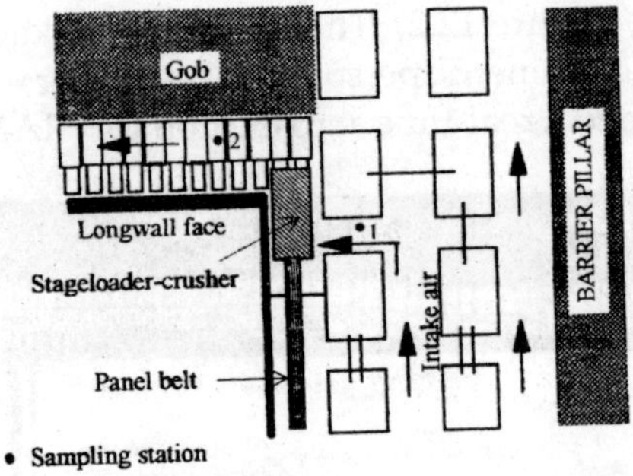

For Mine 1, Mine 2, and Mine 3.

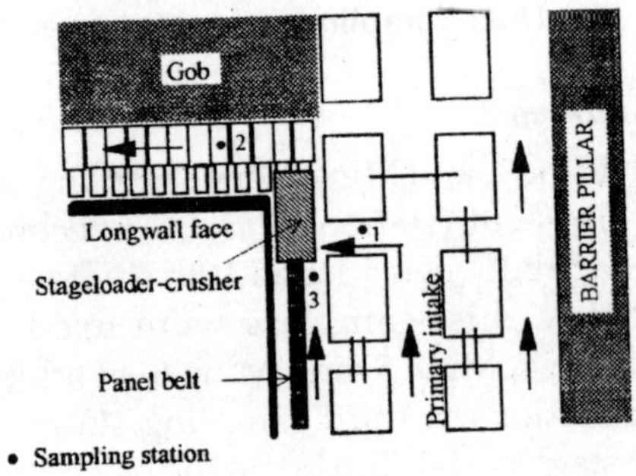

For Mine 4, Mine 5, and Mine 6.

Fig. 12.3: Layout of sampling stations in longwall panels

As per method - 1,

$$R = 0.8\,(C_c - C_o)\,(Q/P) \qquad (1)$$

Where,

C_c = Average ARD concentration in mg/m^3 calculated for the period the shearer cuts a segment of the face upwind of the RAM unit;

C_o = Average ARD concentration in mg/m^3 at the same RAM unit when the shearer is downwind of the RAM during the entire period of experiment;

Q = Air quantity in m^3/s flowing through that sampling location; and

P = Average production rate in tonne/s in the segment concerned.

As per Method - 2,

$$R = 0.8\,(C_{L2} - C_{L1})\,[(Q_{L1} + Q_{L2})/2P] \quad \text{......................(2)}$$

Where,

C_{L1} = Average ARD concentration in mg/m³ at a RAM unit upwind of the segement being cut by shearer;

C_{L2} = Average ARD concentration in mg/m³ at other RAM unit downwind in the segment being cut;

Q_{L1}, Q_{L2} = Air quantities at the two RAM locations in m³/s; and

P = Production rate in tonne/s.

The average ARD concentrations used in the above equations are calculated from the dust concentration profile recorded by RAM excluding the portions during which the shearer was idle. The length of the segment considered are at least 25 shields wide.

Contribution of Outbye Sources

The contributions of stageloader-crusher (C_s), belt entry (C_b), and primary intake (C_i), all in mg/m³, in contaminating the face intake air are given by the following equations:

$$C_s = C_1 - [(C_1Q_1 + C_3Q_3) / (Q_1 + Q_3)]$$

$$C_b = C_3Q_3 / (Q_1 + Q_3)$$

$$C_i = C_1Q_1 / (Q_1 + Q_3)$$

Where,

C_1, C_2 and C_3 are the average ARD concentrations at stations 1, 2, and 3, respectively; and

Q_1, Q_2 and Q_3 are the air quantities flowing through stations 1, 2 and 3, respectively.

On having determined the contribution of the stageloader-crusher in contaminating the face air, the production (P tonnes) during a given time period (t seconds), and total air quantity (Q m³/s), the dust generation rate of the same (R_{stl} mg/tonne) can be calculated as

$$R_{stl} = Cs\,Q\,t / P$$

Results

Dust Generation Rate of the Shearer

With the required average ARD concentrations at the sampling stations, air quantity data and production rate for the equations (1) and (2), the average ARD generation rates of shearer for each experiment in longwall panels of each mine were calculated. The summarised results are presented in Table 12.2. The histograms shown in Figure 12.4 indicate the distribution of dust generaion rates of shearer.

Table 12.2: Summarised results of dust generation rates of shearer

Mine	*Method*	*No. of observations*	*Dust generation rate (mg/tonne)*			*Variance (mg/tonne)²*	*Coefficient of variation*
			Min.	*Max.*	*Average*		
Mine 1	Method-1	55	26	939	217	48130	1.01
	Method-2	54	5	1172	231	68218	1.13
Mine 2	Method-1	88	6	706	113	17711	1.18
	Method-2	76	6	827	133	31500	1.33
Mine 3	Method-1	61	6	373	127	6236	0.62
	Method-2	55	5	284	115	4524	0.58

The results, thus, indicate that although the dust generation rate of shearer varies within a wide range, most of the times it is below 200 mg/tonne. Very high ARD generation rate of the shearer as observed in particular length segments of the face is attributed to the local conditions such as cutting into roof rocks in the region. It is also observed that generation of respirable dust by the shearer is normally higher for the head-to-tail cutting pass in comparison to the tail-to-head cutting pass. This is primarly because the headgate cutting drum during head-to-tail pass cuts along the floor in an exposed position to the ventilating air current directly, whereas during the tail-to-head cutting pass the headgate cutting drum is shielded from the ventilating air by the coalface making it more difficult for the dust from being entrained in the air stream.

Contrbution of Outbye Sources in Contaminating Face Air

Stageloader-Crusher

The airborne respirable dust concentration and air quantity measured at station 1 and station 2 in Mine 1, Mine 2 and Mine 3 are shown in the Table 12.3. The increase in the concentration from station 1 to station 2 is due to the dust generated from the stageloader-crusher and headgate cut-out operations. The mean contribution of the stageloader-crusher and headgate cut-out in Mine 1, Mine 2 and Mine 3 were 0.37 mg/m³, 1.31 mg/m³ and 1.34 mg/m³, respectively. Clearly, the stageloader-crusher and headgate cut-out operation in combination continues to be a major source of contamination at the beginning of the longwall face.

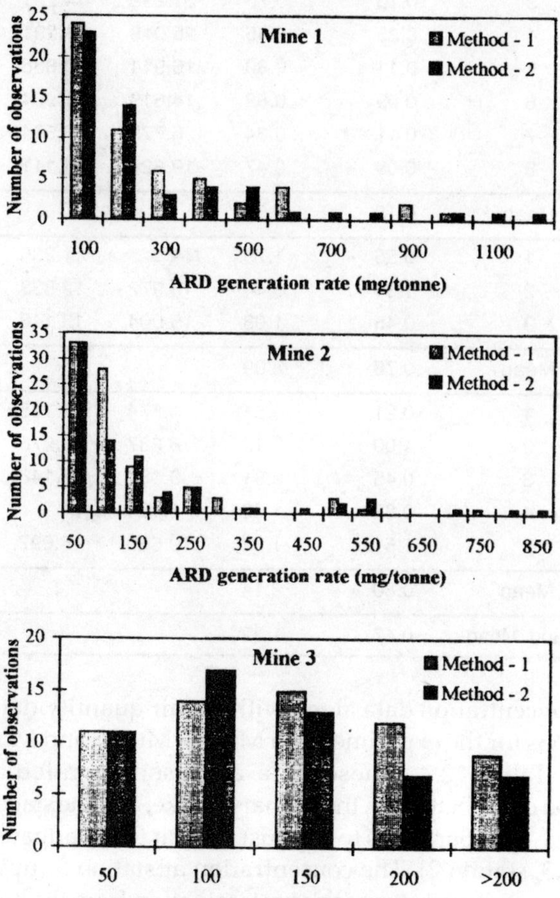

Fig. 12.4: Distribution of dust generation rates of shearer

Table 12.3: ARD concentration and air quantity at station 1 and station 2

Mine	Experiment No.	ARD concentration mg/m^3		Air quantity m^3/s		C_2-C_1 mg/m^3
		Station 1 (C_1)	Station 2 (C_2)	Station 1	Station 2	
1	1	0.38	0.42	25.485	11.706	0.04
	2	0.07	0.32	NA	14.358	0.25
	3	0.13	0.75	21.252	14.231	0.62
	4	0.25	0.45	25.018	11.291	0.20
	5	0.11	0.80	16.914	9.836	0.69
	6	0.09	0.62	14.619	11.964	0.53
	7	0.11	0.34	26.775	6.584	0.23
	8	0.09	0.47	19.326	14.111	0.38
	Mean	0.15	0.52			0.37
2	1	0.56	1.75	NA	11.235	1.19
	2	1.34	3.44	18.972	12.833	2.10
	3	0.45	1.08	15.004	13.148	0.63
	Mean	0.78	2.09			1.31
3	1	0.21	1.51	3.874	4.165	1.30
	2	1.00	3.13	6.387	3.276	2.13
	3	0.45	2.91	8.187	7.148	2.46
	4	0.83	1.32	5.124	NA	0.49
	5	1.51	1.82	10.660	5.697	0.31
	Mean	0.80	2.14			1.34
	Grand Mean	0.47	1.32			0.85

The concentration data along with the air quantity data at the three stations for the experiments in Mine 4, Mine 5 and Mine 6 are shown in Table 12.4. These data are used to calculate the contribution of the belt entry, the primary intake, and the stageloader-crusher/cut-out operations to the dust level at the headgate station (Figure 12.3, station 2). The concentration at station 3 (upwind of stageloader-crusher) is due to the contamination from the belt entry. The concentration at station 1 (intake cross-cut) is due to the contamination by sources upwind in the primary intake. The concentration at station 2 (headgate station) is due to the concentrations at station 3, station 1 and the addition of dust by the stageloader-crusher and headgate cut-out operations.

Table 12.4: ARD concentration and air quantity at the three sampling stations in longwall panels

Mine	*Experiment No.*	*ARD concentration mg/m³*			*Air quantity m³/s*		
		Station 3	*Station 1*	*Station 2*	*Station 3*	*Station 1*	*Station 2*
4	1	1.120	0.220	0.870	7.438	9.713	14.161
	2	0.700	1.600	1.620	7.810	15.038	19.040
	3	0.916	0.280	1.460	8.814	17.481	14.704
5	1	0.437	0.302	0.481	15.223	NA	14.498
6	1	0.527	0.242	1.021	15.223	NA	26.051
	2	0.541	0.348	0.960	16.915	7.646	22.895

The contribution of the stageloader-crusher and panel belt in contaminating face intake air is shown in Table 12.5. The contribution of the stageloader-crusher/headgate cut-out operations to the face intake dust level ranged from 0.26 to 0.97 mg/m³ and as a per cent of the dust level at station 2, it is between 20 per cent to 66 per cent and between 13 per cent to 48 per cent of the permissible exposure limit of 2 mg/m³. In Mine 4, data was also collected on the face concentrations. It was observed that the contribution of the stageloader-crusher/headgate cut-out operations ranged from 13 per cent to 56 per cent of the average face concentration.

Table 12.5: Contribution of stageloader-crusher and panel belt in contaminating intake air

Mine	*Experiment No.*	*Contribution (%)*			*Contribution (%)*			*Contribution (% of PEL*)*		
		Belt entry	*Primary intake*	*Stage-loader*	*Belt entry*	*Primary intake*	*Stage-loader*	*Belt entry*	*Primary intake*	*Stage-loader*
4	1	56%	14%	30%	0.49	0.12	0.26	24%	6%	13%
	2	15%	65%	20%	0.24	1.05	0.33	12%	53%	15%
	3	21%	13%	66%	0.31	0.19	0.97	15%	9%	48%
6	2	39%	11%	50%	0.37	0.11	0.48	19%	5%	24%
	Mean	33%	26%	42%	0.35	0.37	0.51	18%	18%	25%

* Permissible exposure limit of 2 mg/m³

During the experiments in Mine 4, data was also collected on the production from the longwall face and a time study was conducted on the shearer operation. Using the RAM data collected at station 3 and station 2 (when the shearer is downwind of station 2) and assuming that the primary intake concentration does not change during these periods, it was possible to determine the contribution of the stageloader-crusher alone to the concentration at station 2. Knowing the production during these periods, total air quantity (sum of air quantity at station 1 and station 3) and the contribution of the stageloader-crusher alone, the dust generation rate of the stageloader-crusher was calculated in terms of mg/tonne. These rates during experiments 1, 2 and 3 were 14, 60 and 47 mg/tonne, respectively. Difference in the cutting rates of the shearer and changes in the air quantity (primary intake and belt air) are some of the reasons for the variation in the dust generation rates.

Panel Belt Entry

In mines 4, 5 and 6, for all the sampling stations along the panel belt entries, the respirable dust concentrations and the air quantities flowing in the entry at these locations were obtained. In all the panels, the belt speed was approximately 3.25 m/s; the air velocities ranged from 0.81 m/s to 2.50 m/s. Air quantity flowing through belt entry ranged between 34 per cent to 69 per cent of the total quantity delivered at the face. The respirable dust mass flow rates at the stations (the product of the air quantity at the station times the average airborne respirable dust concentration at the station) is used to determine the rate of contamination of the air flow along the belt entry. The ARD concentration, the air quantity and ARD mass flow rates in the belt entries of longwall panels are shown in Figure 12.5.

In all the longwall panels, the air quantity increased as the air moved towards the face due to leakage from the primary intake. The respirable dust concentration either remained unchanged or increased from the head of the belt towards the face. As a result, the average respirable dust mass flow rate increased towards the face. The incremental mass flow rate of respirable dust in the belt entry air for every 305 m (1000 ft) length has been calculated for all the experiments (Table 12.6). Increasing mass flow rate of respirable dust in the belt entry air flowing towards the face can be due to dust entrained from the belt, walking or other movements in the belt entry.

As the floor was generally wet, contribution of walking and other movements can be considered negligible.

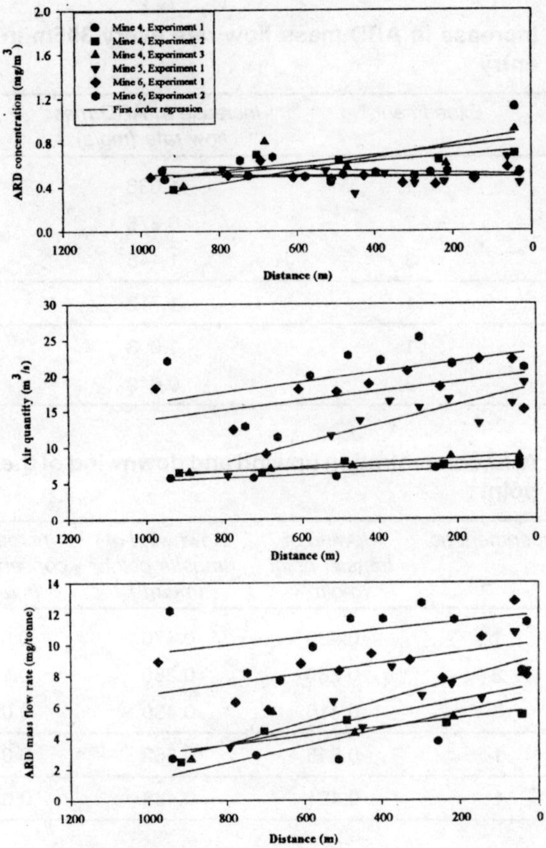

Fig. 12.5: ARD concentration, air quantity and ARD mass flow rate along the panel belt entry

Panel-to-main Belt Transfer Point

The respirable dust concentrations at the sampling stations established to quantify the dust generated by the panel-to-main belt transfer point in mines 4, 5 and 6 are shown in Table 12.7. Dust generated at this outbye source ranged from 0.017 mg/m^3 to 0.1 mg/m^3. In Mine 4, the air in belt entries travel all the way from the surface through the main belts to panel belts. This may be the reason for the comparatively higher rate of contamination in Mine 4.

Table 12.6: Increase in ARD mass flow rate every 305m in the belt entry

Mine	*Experiment No.*	*Increase in ARD mass flow rate (mg/s)*
4	1	1.638
	2	0.875
	3	1.445
5	1	1.713
6	1	1.003
	2	0.612

Table 12.7: ARD concentration upwind and downwind of the transfer point

Mine	*Experiment no.*	*Upwind of transfer point (mg/m³)*	*Downwind of transfer point (mg/m³)*	*Increase in concentration (mg/m³)*
4	1	0.430	0.470	0.040
	2	0.280	0.380	0.100
	3	0.410	0.430	0.020
5	1	0.545	0.562	0.017
6	1	0.461	0.488	0.027

Cumulative Effects

In longwall production panels, the cumulative effect of all the outbye sources can be quite substantial. The RAM dust profile from one of the experiments in Mine 4 is shown in Figure 12.6. The RAM was mounted on shield 8. There were 162 shields at the face and, therefore, the shearer was upwind of the RAM for less than 5 per cent of the operating time. Yet, the RAM profile indicates significant peaks in the concentration profile when the shearer is downwind the RAM, presumably from all sources outbye the face.

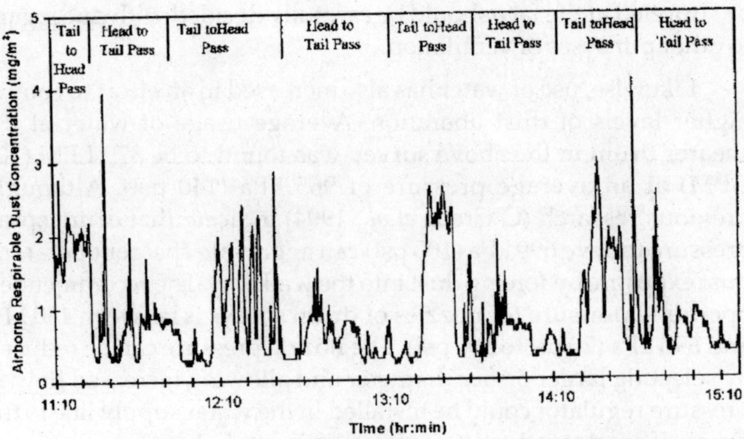

Fig. 12.6: ARD concentration profile recorded by RAM at shield 8 in the longwall face in mine 4, experiment 1

Dust Control Techniques

Control of Dust from Sources at the Face

The amount of dust generated by the shearer is affected by some machine parameters such as, pick size and spacing, drum rotation speed, bit sharpness and machine advance rate, etc. Lower drum rotation speed with increased pick penetration removes larger coal fragments leading to significant reduction in dust generation by the shearer. Wider web cutting also reduces the amount of dust generated per tonne of coal mined. Since it has been found that cutting into the roof and floor rocks also produces a higher amount of dust, this can be avoided by proper horizon control during cutting.

Airflow and water application are the primary dust control measures. Increased quantity of airflow (upto 50m^3/s) is being supplied at longwall faces as revealed during a survey by US Bureau of Mines (Colinet *et al.*, 1997). Due to dilution effect, increased airflow will reduce the concentration of dust in the ambient air at the face. Past research has indicated that air velocities greater than 2.0 m/s should be utilised for improved dust control. Again at higher air velocities, there may be chances of entrainment of dust resulting in increased concentration. Thus, air velocities at the face has important bearing on the dust level at the face and hence the quantity of air to be supplied at the face should be carefully decided with due regard to other purposes of ventilation.

Likewise, use of water has also increased in an effort to control higher levels of dust liberation. Average usage of water at the shearer drum in the above survey was found to be 379 LPM (100 GPM) at an average pressure of 965 kPa (140 psi). Although, previous research (O'Green *et al.*, 1994) indicate that drum spray pressures above 690 kPa (100 psi) can aggravate shearer operators' dust exposure by forcing dust into the walkway. The recommended operating pressure for nozzles of drum sprays is between 480 kPa and 690 kPa (70 psi to 100 psi). The nozzle pressure can be reduced by selecting larger orifice diameter and allowing increased flow. A pressure regulator could be installed in the water supply line to the drums if greater reduction in pressure is needed.

An air-moving shearer-clearer type of external spray system on the body of the shearer effectively reduces the exposure of shearer operator. This spray system is designed to confine dust generated at the shearer drums near the face such that the ventilating air current can carry away the dust from the shearer operators before dust migrates into the walkway.

Use of radio remote control unit can further help reducing the dust exposure of shearer operators by allowing them to position themselves in the area protected by the shearer-clearer sprays.

To reduce the high instantaneous exposure of the shearer operator during the headgate cut-out operation, a curtain in the headgate close to the beginning of the face can be very effective. To prevent the entrainment of dust from the coal being transported over the AFC, low quantity water spray over panline at regular intervals can be used.

Controlling Dust from Sources Outbye the Face

Based on the previous investigations, there can be several methods to control the dust generated by the outbye sources. These methods are as follows:

Homotropal Ventilation

The most effective method for controlling intake dust is homotropal ventilation. which routes air in the direction of coal transportation along the face, *i.e.* tailgate-to-headgate. This places all the outbye sources downstream of the face workers, eliminating their dust exposure from these sources. Intake dust levels in the walkway can be reduced by as much as 90 per cent (Organiscak *et al.*, 1986). However, for adopting the homotropal ventilation system, the tailgate entries must be kept in good conditions.

Control of Dust Generated by Stageloader-Crusher

Significant dust reduction can be achieved with water sprays located strategically on an enclosed stageloader-crusher, or with water-powered scrubber that cleans the air inside it.

A study by Organiscak *et al.* (1986) indicated that by enclosing the stageloader-crusher and locating the water sprays strategically (four hollow cone sprays in the crusher, spray bar with two hollow cone sprays directed at the intake of the crusher, spray bar with three hollow cone sprays directed down on the chain conveyor on the discharge side of the crusher, and a spray bar with three hollow cone sprays directed on and with coal movement on the conveyor belt immediately after the stageloader-dump point)) reduced the dust concentrations considerably at the headgate operator (79 per cent), stageloader (74 per cent) and at support 20 (41 per cent); and modest dust reductions in the belt entry (17 per cent). Further, 30 per cent additional improvement was noticed at the headgate operator's position and at support 20 on increasing the water flow from 76 LPM (20 GPM) to 114 LPM (30 GPM).

Another means of controlling dust from the stageloader-crusher is the water-powered scrubber that cleans the air inside an encolsed stageloader-crusher and discharges the clean air into the mine environment (Figure 12.7). In this system, the scrubber is located as close as practical to the crusher discharge and exhausting the clean air in the direction of coal movement. With approximately 34 LPM (9 GPM) water at 3450 kPa (500 psi), an airflow of about $1 m^3/s$ (2000 cfm) can be moved through the scrubber unit resulting in significant reductions in dust levels at the headgate operator's position and in the intake air entering the longwall face.

Control of Dust from Panel Belt

During outbye transport, the water on coal may evaporate making rewetting necessary at intervals which can reduce the dust generated during the transport. A low quantity flat-fan sprays or full-cone sprays (applying between 4-5 LPM water at 345 kPa or more) can be used for rewetting the coal over the panel belt. Such a system can be used in association with a sensor to sense the coal flow such that the sprays are operative only during the coal transport.

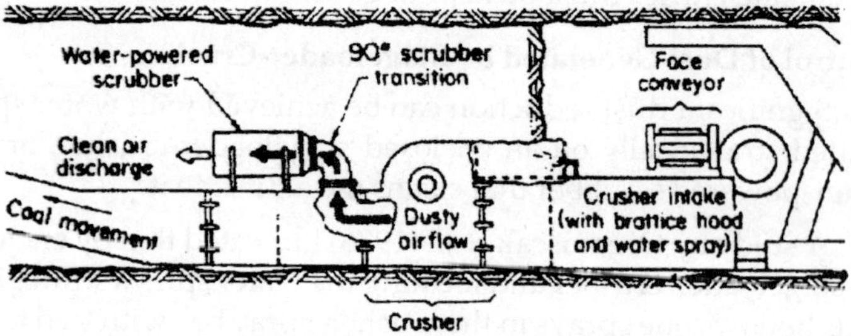

Fig. 12.7: Enclosed stageloader-crusher with water-powered scrubber

Belt maintenance and belt cleaning also have a bearing on the dust level in the panel belt entry. Missing rollers, belt slippage, and worn belts can create dust problems. The top and bottom of return belt should be cleaned with spring loaded or counter-weighted scrappers to prevent material adhering to the return belt from being pulverised by the tail-end pulley. A low-quantity water spray can also be used to moisten the belt slightly. Other means of cleaning the return belt are a motor-driven rotary wire brush that cleans the conveying side by rotating opposite belt travel, and a full cone water spray directed at the non-conveying side followed by a piece of carpeting (belt wiper) positioned across the belt width. The water spray wets the fines and the carpeting removes them from the non-conveying side.

The floor of the belt entry and intake airway should be kept wet to reduce the intake dust level. Plain water dries out rapidly in the ventilating air making repeated application necessary. Addition of calcium chloride or other hygroscopic salt will be able to maintain the moisture content at the desired level. Dust level in the primary intake can also be controlled by limiting the construction activities, haulage of supplies, and roadway grading, etc. during the production shift.

Discussion

This study has quantified the dust generation rate of shearer at longwall faces and contribution of outbye sources (stageloader-crusher, panel belt and panel-to-main belt transfer point) to the contamination of face intake air.

Dust generation rate of the shearer depends on machine, operational, and geo-mining parameters and their complex relationships. It is observed from this study that though the dust generation rate of shearer varied within wide range, most of the times it was less than 200 mg/tonne. Higher rates of dust generation were observed due to cutting into roof rock.

The contribution of the stageloader-crusher ranged from 20 per cent to 66 per cent of the dust level of face intake air. The contribution of the panel belt entry also ranged between 15 per cent to 56 per cent. The ARD concentration in the panel belt entry either remained unchanged or increased and the ARD mass flow rate was found to be increasing as the air moved towards the face. Increase in the ARD concentration along the panel belt entry was perhaps due to entrainment phenomena since the floor appeared to be generally wet. The maximum relative air velocity during the experiments was 5.75 m/s with respect to coal over belt. The situation can be different for higher relative air velocities. Dust generated at the panel-to-main belt transfer point ranged from 0.017 mg/m^3 to 0.1 mg/m^3. Although each outbye source of airborne respirable dust individually may not appear to be significant, their collective effect on the exposure of face workers can be matter of concern since the face workers remain downwind of them throughout the shift.

Suitable engineering dust control measures at each source should be adopted to reduce that dust exposure of longwall face workers and to comply with the mandated dust level as well. A few of them have been discussed in this paper which can be easily adopted for effective control. Further, the findings of the study can be helpful in ventilation planning with a view to reduce the exposure of the longwall face workers.

References

Colinet *et al.* (1997). "Status of dust control technology on US longwalls". Proceedings of the 6th International Mine Ventilation Congress.

Haney, R.A. (1996). "Effect of belt air on dust levels in underground coal mines". Applied Occupational and Environmental Hygiene, Vol. 11, No. 7, pp. 826-829.

Jankowski, R.A. and Organiscak, J.A. (1996). "An overview of research experience as it relates to shearer dust control in the United States", Proceedings of the International Scientific and Technical Conference on Respirable Dust Hazard Control in the World Mining Industry, KOMAG Mining Mechanization Centre, Gliwice, Poland, pp. 5-10.

Jankowski, R.A. and Organiscak, J.A. (1996). "US longawall practices for controlling respirable dust sources outby the shearing machine", Proceedings of the International Scientific and Technical Conference on Respirable Dust Hazard Control in the World Mining Industry, KOMAG Mining Mechanization Centre, Gliwice, Poland, pp. 19-25.

Jankowski, R.A. and Organiscak, J.A. (1983). "Dust sources and controls on the six U.S. longwall faces having the most difficulty complying with dust standards", Bureau of Mines Information Circular, *IC 8957*, U.S. Department of Interior, Washington D.C., 19 pp.

Jankowski, R.A. and Hake, J. (1989). "Dust sources and controls for high production longwall faces", *Proceedings of Longwall USA,* Pittsburgh, pp. 117-132.

Jankowski, R.A., Jayaraman, N.I. and Organiscak, J.A. (1990). "Dust sources and controls on high tonnage longwall faces", SME Preprint No. 90-73, Society for Mining, Metallurgy, and Exploration, Inc., Littleton, CO.

Organiscak, J.A., Jankowski, R.A. and Kelly, J.A. (1986). "Dust controls to imrpove quality of longwall intake air". Information Circular 9114, United States Department of Interior, Bureau of Mines, 8 pp.

Organiscak, J.A., Jankowski, R.A. and Kelly, J.A. (1986). "Dust controls to improve quality of longwall intake air". Bureau of Mines Information Circular, *IC 9114*, U.S. Department of Interior, Washington D.C., 8 p.

O'Green, J.E., McNider, E.T., Gallick, J.M., Jarrett, R.J. and Jankowski, R.A. (1994). "An overview of general operating experience as it relates to longwall dust control". *Proceedings of Longwall USA*, Pittsburgh, pp. 221-232.

Paschedag and Martin (1999). "DBT Reshapes Longwall Mining Worldwide". Coal Age, October 1999, pp. 16-24.

13

Impact of Dishwashing Activity on Health of Housewives

Nutan, R. Biradar
Department of Family Resource Management,
University of Agricultural Sciences,
Dharwad - 580005, Karnataka

Introduction

In the changing world women are exposed to the various gadgets of modern technology, which will help them to carry out many household work within a short period of time and least expenditure of energy. But under Indian condition majority of household works need to be attended by the housewives and dishwashing is one among them. It is being carried out in various postures and body positions. These different body positions form a combination of static and dynamic mascular efforts which cause common problem of backache, consequenlty housewives suffer from tiredness and exhaustion which adversely affect the performance of the household work. Dishwashing is an activity which causes a lot of physical stress to the human body in terms of energy expenditure. The energy expenditure is the objective measurement of physical cost of work. Hanson and Jones (1970) reported that the hear trate differed significantly among many postures but not between the conditions of the task. Pain and discomfort are the feed back information from the body to alert disease, damage or the body limitations. The perceived discomfort can be recorded in terms of pain felt in various parts of the body. Pain is the indicator of discomfort (Corlett and Bishop, 1976).

Dishwashing though an important household acitivity, has yet remained unattempted to know more scientifically, so the present study was undertaken to know the postural adoption, changes in posture, to measure the perceived pain and discomfort while working and also to assess the effect of the activity on heart rate, energy expenditure and angles of body bend while cleaning the dishes at different sink heights and to suggest the optimum sink height for dishwashing activity.

Materials and Methods

The study was conducted in Dharwad city of Northern Karnataka. The purposive sampling techniques was adopted to select 90 respondents. The postural changes during the process of dishwashing activity was recorded with the help of postural observation chart.

Procedure to Record the Perceived Pain and Discomfort

The perceived pain and discomfort was recorded with two interval namely, 20 minutes after starting the work and at the end of the work. To record the pain and discomfort the body diagram flash card was used (Fig. 13.1). It consist of various parts of the body numbered separately. After 10 minutes of starting the work the respondents were asked to indicate the site of pain and discomfort on the diagram. This process was continued until no pain and discomfort responses were recorded in recording sheet.

For laboratory study ten female postgraduate students were selected based on the elbow height. The activity was carried out in standing posture. The body positions studied for dishwashing were, sink at elbow height, sink at 10 cm less than elbow height, sink at 20 cm less than elbow height and at laboratory sink height of 76.5 cm.

Method for Estimation of Energy Expenditure

The energy expenditure during the work is estimated with the help of average working pulse rate, by using the following formula. It is expressed in kilo calories per minute. The estimated energy expenditure in kilo calories per minute was worked out by using the following formula:

$$\text{Estimated energy expenditure} = 0.0359 \times \text{WHR} - 1.398 \times X/50$$

Where,

0.0359 = Constant

WHR = Working pulse rate

X = is observed weight

50 = constant (mean body weight)

1.398 = constant

(Source: Dr. Suhasini Rao, 1987).

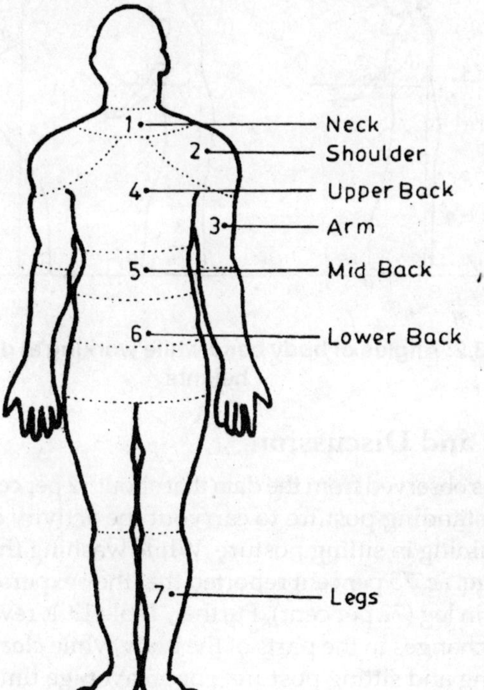

Fig. 13.1: Body map

Measurement of Body Angles

The angles of body bend were measured from the photographs of four dynamic postures namely, sink at elbow height, 10 cm less than elbow height, 20 cm less than elbow height and at laboratory sink height of 76.5 cm. For measuring the angles, land marks on the body were considered, namely, acromial and slylian, (Landmarks on body were marked by fixing an adhesive tape with a 'X'). Angle of back banding, neck and head tilting and elbow extension were measured with a known reference point namely, back, neck head and elbow in normal position. In each posture these angels were measured with the help of a protractor (Fig. 13.2).

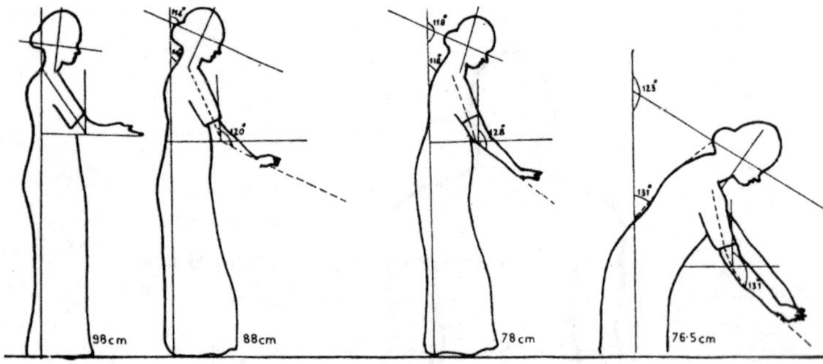

Fig. 13.2: Angles of body bend while working at different sink heights

Results and Discussion

It was observed from the data that about 52 per cent of the sample adopted standing posture to carryout the activity of dishwashing and remaining in sitting posture. While washing the dishes higher percentage, *i.e.* 75 per cent reported that they experienced backache and pain in leg (72 per cent). Further, Table 13.1, revealed the mean postural changes in the parts of the body while cleaning the dishes in standing and sitting posture. For an average time of 25 minutes the frequent changes in the posture of back and neck *i.e.* forward bending was noticed to a greater extent (319.00 and 306.00 times, respectively) when compared to other parts. However, frequency of changes for the above mentioned parts, *i.e.*, forward bending was still higher in sitting posture, *i.e.* 381.00 times for neck and 371.00 time for back. Thus, change in the posture was more in sitting than standing, there by it indicates that sitting posture may not be ideal for the activity. Also, it may be due to improper working condition.

Table 13.1: Mean postural changes in the parts of the body while cleaning the dishes in standing and sitting postures

Postural change	*Body Parts*							
	Neck		*Back*		*Arms*		*Leg*	
	Standing	*Sitting*	*Standing*	*Sitting*	*Standing*	*Sitting*	*Standing*	*Sitting*
Straight	142.00	101.00	115.00	83.00	-	-	-	-
Forward bending	306.00	381.00	319.00	371.00	-	-	-	-
Rotated	66.00	55.00	68.00	40.00	-	-	-	-
Right angle	-	-	-	-	33.00	28.00	-	-
Flextion	-	-	-	-	59.00	56.00	-	-
Extension	-	-	-	-	308.00	347.00	-	-
Both leg straight	-	-	-	-	-	-	217.00	-
One leg bent forward	-	-	-	-	-	-	155.00	-

A glance at Table 13.2 indicates the perceived pain and discomfort expressed by the respondents while carrying out the activity. It was found that, perceived pain and discomfort in standing posture soon after 10 minutes of starting the work and at end of the work was found to be slight. The observed overall percentage for slight and more pain and discomfort after 10 minutes of starting the work and at end of the work was 20.43, 9.56 per cent and 23.48, 14.35 per cent, respectively. However, slight pain and discomfort was reported in upper parts of the body namely, shoulder (16.30 per cent), neck (14.31 per cent), arm (11.95 per cent), and lower parts namely back (35.86 per cent) and leg (31.52 per cent) soon after 10 minutes of starting the work and this was reported by higher percentage of the respondents. Mean perceived pain and discomfort for back and leg was comparatively high. An average time spent for dishwashing activity was 25 minutes. While, data on perceived pain and discomfort in sitting posture brought to the focus that more pain and discomfort after 10 minutes of the work was not perceived and reported by the selected sample. According to higher percentage of the respondents, slight pain and discomfort was felt at the end of the work. The recorded overall mean percentage of perceived pain and discomfort was 46.36 during end of the work. The percentage for perceived pain and discomfort for all the parts of the body at end of the work in sitting was comparatively high than from standing posture. The mean perceived pain and discomfort for shoulder was found to be high *i.e.*, 28.40 per cent followed by arm (19.31 per cent) and neck (14.77 per cent). Between back and leg, back revealed higher percentage (50.00 per cent) than leg (44.31 per cent).

Further, association between posture and pain and discomfort was studied at applying chi-square (Table 13.3). Between the posture *viz.*, standing and sitting the respondents expressed pain and discomfort comparatively high in the back after 10 minutes of starting the work in standing posture and it was found to be slight. This association was found to bc significant at 0.05 level. However, at end of the work both slight pain and discomfort and more pain and discomfort in the back and leg was high both in sitting and standing posture, but, the association was not significant.

It is observed that an overall mean energy expenditure was more while cleaning dishes at laboratory sink height (2.14 kcal per minute) followed by sink at 20 cm less than elbow height (2.00 kcal per minute) and sink at elbow height level (1.81 kcal per minute) cleaning dishes at sink height of 10 cm less than elbow height was found more suitable as it consumed less energy, *i.e.*, 1.72 kcal per minute when compared to other sink heights (Table 13.4).

The pulse rate was recorded thrice with an interval of one minute and it was observed that, sink at 10 cm less than elbow height was found to be more suitable as an average working pulse rate was found to be less (91.99 beats/min) when compared to all the other sink heights.

The angles of the body bend (Table 13.5) discloses that, the sink at 10 cm less than elbow height was found to be the better when compared to other sink heights, as the forward bending of back, neck, head and arm extension was less for all the subjects when compared to other two sink heights *i.e.*, sink at 20 cm less elbow height and at laboratory sink height. This could be due to too much of bending forward, and this may lead to compression of organs and thereby increasing the pulse rate (Fig. 13.2). Dhilion (1982) also pointed out that, various body positions accounted for variations in the heart rate.

Table 13.2: Perceived pain and discomfort while cleaning the dishes by the selected housewives

Posture and frequency	*Time of measurement*	*Perceived pain and discomfort*											
		Slight pain and discomfort						*More pain and discomfort*					
		N	*S*	*A*	*B*	*L*	*Overall percentage of perceived pain and discomfort*	*N*	*S*	*A*	*B*	*L*	*Overall percentage of perceived pain and discomfort*
Standing (51.11)	10 minutes after after starting the work	7.00 (15.21)	11.00 (22.00)	7.00 (15.21)	14.00 (30.29)	8.00 (17.29)	9.40 (120.43)	2.00 (4.34)	3.00 (6.52)	4.00 (8.54)	7.00 (5.21)	5.00 (13.00)	4.40 (9.56)
	End of the work	6.00 (13.04)	4.00 (8.69)	4.00 (8.69)	19.00 (41.30)	21.00 (45.65)	10.80 (23.48)	4.00 (8.69)	4.00 (8.69)	4.00 (8.69)	11.00 (23.91)	12.00 (23.73)	6.60 (14.35)
Mean percentage of perceived pain and discomfort		6.50 (14.13)	7.50 (16.30)	5.50 (11.95)	16.50 (35.86)	14.50 (31.52)	—	3.00 (6.52)	3.50 (7.60)	4.00 (8.69)	9.00 (19.56)	3.00 (17.39)	—
Sitting (48.69)	10 minutes after starting the work	3.00 (6.81)	7.00 (15.90)	7.00 (15.90)	10.00 (22.72)	8.00 (18.18)	7.00 (15.90)	—	—	—	—	—	—
	End of the work	9.00 (20.45)	18.00 (40.90)	10.00 (22.72)	34.00 (77.27)	31.00 (70.45)	20.40 (46.66)	6.00 (13.63)	3.00 (6.81)	4.00 (9.09)	6.00 (13.63)	5.00 (11.36)	4.80 (10.91)
Mean percentage of perceived pain and discomfort		6.50 (14.77)	12.50 (28.40)	8.50 (19.31)	22.00 (50.00)	19.50 (44.31)	—	6.00 (13.63)	3.00 (6.81)	4.00 (9.09)	6.00 (13.63)	5.00 (11.36)	4.80 (10.91)

Note: Figure in parantheses indicate percentage N = Neck, S= Shoulder, A = Arm, B = Back, L = Leg

Table 13.3: Association between posture and pain and discomfort

Posture	10 minutes after starting the work						End of the work											
	Slight pain and discomfort						Slight pain and discomfort						More pain and discomfort					
	N	S	A	B	L	Total	N	S	A	B	L	Total	N	S	A	B	L	Total
Standing	7.00	11.00	7.00	14.00	8.00	47.00	6.00	4.00	4.00	19.00	21.00	54.00	4.00	4.00	4.00	11.00	10.00	33.00
	(5.73)	(10.31)	(8.02)	(13.76)	(9.18)	(57.31) (100)	(5.20)	(7.61)	(4.85)	(18.34)	(18.00)	(34.60)	(5.79)	(4.05)	(4.63)	(9.84)	(8.69)	(57.90)
Sitting	3.00	7.00	7.00	10.00	8.00	35.00	9.00	18.00	10.00	34.00	31.00	102.00	6.00	3.00	4.00	6.00	5.00	24.00
	(4.27)	(7.69)	(5.98)	(10.24)	(5.82)	(42.69) (100)	(9.80)	(14.39)	(9.15)	(34.65)	(34.00)	(65.40)	(4.21)	(2.95)	(3.37)	(7.16)	(6.31)	(42.10)
Total	10.00	18.00	14.00	24.00	16.00	82.00	15.00	22.00	14.00	53.00	52.00	156.00	10.00	7.00	8.00	17.00	15.00	57.00
	(12.20)	(21.95)	(17.07)	(29.27)	(19.51)	(100)	(9.61)	(14.10)	(8.98)	(33.98)	(33.33)	(100.00)	(17.54)	(12.29)	(14.03)	(29.83)	(26.31)	(100.00)
X^2 value	11.588*						3.840 NS						2.313 NS					

Note: Figure in parantheses indicate percentage
NS = Non significant
* = Singificant at 5 per cent level

N = Neck, S = Shoulder, A = Arm, B = Back, L = Leg

Table 13.4: Energy expenditure while cleaning the dishes

Sl.No.	Elbow height	Body weight	Sink at elbow height	Sink at 10 cm less than elbow ht.	Sink at 20 cm less than elbow ht.	Laboratory sink (76.5 cm)
			Energy expenditure in Kcal	Energy expenditure in Kcal	Energy expenditure in Kcal	Energy expenditure in Kcal
1.	95	47	1.50	1.48	1.75	2.03
2.	95	45	1.77	1.63	1.87	2.05
3.	96	50	1.86	1.66	2.33	2.40
4.	97	40	1.75	1.71	1.86	1.95
5.	97	45	1.61	1.61	1.87	2.08
6.	97	49	1.94	1.62	2.04	2.33
7.	98	41	1.67	1.65	1.84	1.86
8.	99	46	2.05	2.01	2.17	2.26
9.	99	47	1.89	1.85	2.08	2.18
10.	100	47	2.10	2.07	2.22	2.27
Mean	97.30	46	1.81	1.72	2.00	2.14

Conclusion

The findings of the study thus conclude that designing of sink for dishwashing activity needs to be given careful thought. It is true that people will adopt to bad work places but it is at personal cost and with lower productivity. Therefore it has a significant association with the workers's discomfort. As indicated by the research findings, sink height of 10 cm less than elbow height was found more suitable because of less energy expenditure (1.72 kcal/min), least pulse rate (91.59 beats/min) and less angle of body bend (102°). Hence, sink at 10 cm less than elbow height can be suggested as optimum sink height for dishwashing activity. As it reduces the pain and discomfort of the house wives (Fig. 13.3).

Table 13.5: Angles of body bend while cleaning the dishes under laboratory condition

Elbow ht.	Body parts											
	Back				Neck and Head				Arm			
	Sink at elbow ht.	Sink at 10 cm less than elbow hg.	Sink at 20 cm less than elbow ht.	lab sink (76.5 cm)	Sink at elbow ht.	Sink at 10 cm less than elbow ht.	Sink at 20 cm less than elbow ht.	lab sink (76.5 cm)	Sink at elbow ht.	Sink at 10 cm less than elbow ht.	Sink at 20 cm less than elbow ht.	lab sink (76.5 cm)
95 cm	Straight (90°)	115° (90+25°)	131° (90+41°)	137° (90+47°)	Straight (90°)	117° (90+27°)	131° (91+41°)	131.5° (90+41.5°)	Right angle (90°)	110.5° (90+20.5°)	121.5° (90+31.5°)	130° (90+40°)
96 cm	Straight (90°)	113° (90+23°)	122° (90+32°)	130° (90+40°)	Straight (90°)	105° (90+15°)	132° (90+42°)	142° (90+52°)	Right angle (90°)	107° (90+17°)	119° (90+29°)	121° (90+31°)
97 cm	Straight (90°)	109° (90+19°)	119° (90+29°)	140° (90+50°)	Straight (90°)	112° (90+22°)	126° (90+36°)	128° (90+38°)	Right angle (90°)	109° (90+19°)	125° (90+35°)	127° (90+37°)
98 cm	Straight (90°)	106° (90+16°)	118° (90+28°)	131° (90+41°)	Straight (90°)	114° (90+24°)	114° (90+24°)	123° (90+33°)	Right angle (90°)	120° (90+30°)	128° (90+38°)	132° (90+42°)
99 cm	Straight (90°)	104) (90+14°)	114.5° (90+24.5°)	132.5° (90+42.5°)	Straight (90°)	102.5° (90+12.5°)	105.5° (90+15.5°)	112° (90+22°)	Right angle (90°)	114° (90+24°)	133° (90+43°)	130° (90+40°)
100 cm	Straight (90°)	104° (90+14°)	120° (90+30°)	130° (90+40°)	Straight (90°)	104° (90+14°)	112° (90+22°)	119° (90+29°)	Right angle (90°)	102° (90+12°)	111° (90+21°)	120° (90+30°)

Note: Sink at elbow height was considered as referring point.

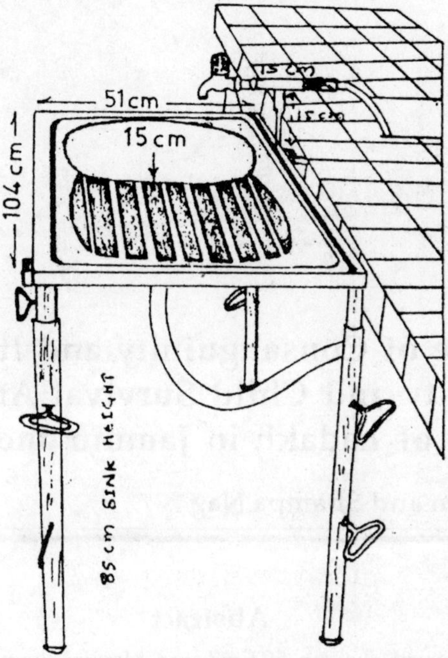

Fig. 13.3: Recommended sink dimensions

References

Corlett, E.N. and Bishop, R.P. (1976). A technique for assessing postural discomfort *Ergonomis*, 19(2): 175-182.

Dhillon, M.K. (1982). Comparative study of physiological costs in different postures of dish washing . *Ph.D. Thesis,* Punjab Agircultural University, Ludhiana.

Hanson, J.A. and Jones, F.P. (1970). Heart rate and small postural changes in man. *Ergonomics*, 13(4): 483-487.

Suhasini Rao, S.J. (1987). Physiological costs and perceived exertion for home and farm activities done by rural women. *Ph.D. Thesis*, SNDT Women's University Bombay, Bombay.

14

Incidence of Consanguinity and Its Effects on Fertility and Child Survival Among Muslims of Ladakh in Jammu and Kashmir

M.K. Bhasin and Shampa Nag

Abstract

In the present chapter, 503 married Muslim women belonging to Balti, Brokpa and Arghun population groups of Ladakh region have been studied. The incidence of consanguinity have been found relatively low at 19.3 per cent, when compared with other muslim population groups of Northern India. The mean coefficient of inbreeding has been estimated as 0.0116 for Ladakh Muslims as a whole. The most prevalent type of consanguineous marriage is that between first cousins. The Ladakh Muslims also show increased fertility and decreased proportion of surviving children in consanguineous than in non-consanguineous marriages, as noticed elsewhere as well.

Introduction

Marriage - a socio-cultural custom, is the basis for procreation and therefore, certain marriage practices of assortative nature deviating from panmixis - especially those between relatives, have been receiving widespread attention from various fields including, genetics, anthropology, sociology, demography. 'Relatives' according to Vogel and Motulsky (1986) are individuals who have certain portion of their genes in common by descent. Marriage between two such individuals who have at least one traceable common ancester is said to be 'consanguineous' and offspring of such mating 'inbred'. The major points of interest in pursuing studies on consanguinity are its effects on fertility, mortality and morbidity and the risk of increased manifestation of deleterious homozygotes which are believed to influence the demographic and genetic structure of population. The consanguineous matings facilitate the recessive genes to manifest their affects and the semi-dominant genes to increase their effects by becoming homozygous. And, knowledge of these aspects is of considerable importance for studying the etiology of various diseases, defects, developmental disorders, genetic load and for genetic counselling.

The earliest attempts to gauge the incidence of consanguineous marriages and its effects, and to measure the degree of inbreeding in a population group were made by Bemiss (1858). Darwin (1875), Pearson (1902), Arner (1908), Wright (1922), Bell (1940), Malecot (1948), Neel *et al.* (1949). Thereafter, the incidence of consanguinity have been reported from Europe, Ireland, America, Kurdistan, Japan, although the levels of inbreeding appeared rather low, with the inbreeding coefficients rarely exceeding 0.01. Relatively higher inbreeding levels have been observed in isolates and isolated communities such as, the Ramah Navaho Indians (Sphuler and Kluckhonn, 1953), Dunkers (Glass, 1954), Dinkas (Roberts, 1956), Susaks (Doliner, 1960) and in some isolated islands as Hosojima in Japan (Ishikuni *et al.*, 1960), in Samaritan isolates (Bonne, 1963), in island of Tristan da Cunha (Bailit *et al.*, 1966) as well as in a large number of muslim communities in Sudan, Iran, Kuwait, Uzbekistan, Egypt and Pakistan (Ahmed, 1979; Naderi, 1979; Alfi *et al.*, 1980; Ginter *et al.*, 1980; Habib and Book, 1983; Hafez *et al.*, 1983; Al-Awadi *et al.*, 1986; Shami *et al.*, 1989).

In India, inhabited by a vast congregation of castes, communities and Tribes, the consanguineous marriages are being practised for long amongst several groups, though many seem to prescribe the same. But, the studies on incidence and effects of consanguinity have been widely attempted only after the publications of Sanghvi (1954) and Sanghvi *et al.* (1956). There exist a few informative reviews of these studies (Mahapatra, 1966; Chakravartti, 1968; Malhotra, 1976; Roychoudhury, 1976a, b). In general, it appears that whereas some population groups prefer consanguineous marriages, many others proscribe these. Therefore, the observations cannot be generalized successfully for India as a whole.

Recently, in a comprehensive regional review on all aspects of consanguineous marriages in the Indian region by Bhasin and Nag (1994) too, similar picture has emerged. This regional appraisal has also shown that consanguineous marriages are practised not only in small isolates or geographically restricted communities, or Scheduled Tribes but also in numerically large caste groups and communities, due to diverse economic and socio-cultural reasons. It has broadly indicated that the states of the southern region of India overwhelmingly prefer and practise consanguinity across all major religious groups and ethnic entities, except Kerala, where comparatively low incidence has been found, owing perhaps to relatively high educational characteristics, and near absence of a major type of consanguineous marriage (prevalent especially in the southern states), those between Uncle-Niece, most probably due to the prevailing matrilineal traditions there till recent times. In the eastern region, many communities of Maharashtra appeared to submit to the Dravidian influence, thereby preferring such marriages fairly frequently. In the other states of western region; and the northern, eastern region, as well as in the Islands, consanguinity in general, seemed to be proscribed or practised occasionally. However, the review also pointed out the incomplete and inconclusive picture regarding the practice of consanguinity as well as inconsistencies of its effects across various regions and populations of India, as seen elsewhere in the world.

In the present article, an attempt has been made to study the incidence of consanguinity and its effects on fertility and child survival among the Muslims of Ladakh region in Jammu and Kashmir - a hitherto unexplored region regarding the practice. The reasons behind the practice have also been researched. An attempt has additionally been made to compare the study populations with other Indian Muslim population groups from available published materials.

Materials and Methods

The northern most State of Jammu and Kashmir in India can be broadly divided into the following three natural regions/ecological zones, taking into consideration the diversified and remarkable variations in the physiographical conditions: (a) Outer Hills, (b) Middle Himalayas, (c) Inner Himalayas. The last region, which more or less synchronizes with the Ladakh region, is situated in the highland crystalline mass of trans-Himalayan rain-shadow zone and mainly comprises of high-altitude mountains, glaciers, narrow, valleys, alluvial fans and deep gorges.

The Ladakh region is the largest division of the state and covers an area of 96,701 sq. km (includes 37,555 sq km. Under the illegal occupation of China), *i.e.*, 70 per cent of the total area of the state. It is situated between 32°15′ and 36° north latitude and 75°15′ and 80°15′ east longitude. The land surface of the Ladakh region can be broadly divided into the upper zone above 4500 m and the lower zone between 2700 m and 4500 m (approximately). All human settlements and settled agriculture are confined to the lower zone. The Ladakh region, comprising of Leh and Kargil districts constitutes the present study area.

The Ladakh region is inhabited by a number of ethnic groups belonging mainly to the two major religious groups, the Buddhists and Muslims. The practice of consanguinity is generally absent among Buddhists, but seen among Muslims. The muslim population groups studied in this context are: Baltis, Brokpas and Arghuns.

Baltis (Scheduled Tribes)

The Baltis inhabit most the Kargil district and parts of Indus valley and Nubra-Shyok valley in the Leh district. They are believed to belong to the predominant ethnic stock of Ladakh region, *i.e.*, having admixture to Mongoloid and Aryan (Dard) elements. It is suggested that they were earlier Buddhists, but long back converted to Islam when mass conversion took place in Baltistan - the original home of these people. Baltis profess Shia faith of Islam, and therefore they are sometimes referred as 'Shias'. They are mainly agriculturists, and horticulture, animal husbandry are sometimes marginally undertaken. Their spoken language is 'Balti', which belongs to the Tibeto - Chinese family and preserves the archaic characters of pronuciation. However, they use Urdu, Persian characters for writing.

Brokpas (Scheduled Tribe)

'Brok' means high pasture ground; and hence 'Brokpa' connotes 'highlander'. Brokpas are muslim Dards and inhabit the valleys along the Drass river and its tributaries in the Kargil district. They are also known as 'Shinna', after their spoken language, which belongs to the Dard group of languages in the non-sanskritic Indo-European family. The profess Sunni faith of Islam and many of their socio-cultural customs are now akin to those of Sunni Kashmiri Muslims inhabiting the adjacent region and also the Baltis. They have not retained much of original customs and rituals after their conversion to Islam long back, unlike their Buddhist counterparts inhabiting the lower Indus valley, who are also known as 'Brog-pa' in the Leh district, but 'Dog-pa' or Dukpa' in the Kargil district. They belong to the Aryan stock and are believed to have come from Dardistan (the Gilgit region) in distant past. It seems that they had been mentioned in ancient chronicles by the Greek authors. The main occupation of the Brokpas is agriculture. Their subsidiary occupation is animal husbandry and negligibly horticultural activities. Many even venture out for jobs.

Arghuns (Community)

Since long, the Ladakh region occupied a nodal position on the central Asian trade route and, therefore, was frequently traversed by merchants from various regions including Kashmir valley and central Asia. Arghuns are the descendants of Muslim merchants of Kashmir valley, Yarkand, and also muslim missionaries. These people married local Buddhist Ladakh women who converted to Islam and eventually settled there. They profess Sunni faith of Islam. Many of their socio-cultural customs bear similarity with those of the Sunni Kashmiri Muslims of the Kashmir valley. They inhabit mainly the Indus valley in Leh district, but a smaller proportion is also found in the Zanskar valley in Kargil district.

Their spoken language is 'Ladakhi' belonging to the Tibeto-Chinese family, but many can converse in Balti, Turkish, Tibetan and Kashmiri languages. However, Urdu characters are mostly used for writing. They are mainly engaged in trade and commerce. Some however, have taken to agriculture with minimal horticultural and animal husbandry activities.

The study data were collected from 503 married Muslim women (primary respondents) using interview schedules and participant observation. Of these, 298 were Baltis, 54 were Brokpas, and 151 were Arghuns. The samples were drawn randomly from both Leh and Kargil districts of the Ladakh region after studying their distribution pattern. Detailed information on demographic, economic and socio-cultural aspects were collected using probling and the life-history approach method. That is, the integrated pregnancy/birth histories of the respondents were reconstructed as a detailed chronological list. In between, various other events in the household or in respondent's life were also fitted in. The collected information was continuously cross-checked from other members of the household, particularly elders, as well as, local para-medical staff, village heads, religious and local leaders, who were more conversant with the local population. Further, if a marriage was reported as consanguineous, the reason (s) were noted and the family pedigree was carefully drawn extending upward to two earlier generations on the side of each spouse in order to determine the existence and type of consanguinity involved. Consanguineous marriages in the present study were classified into: (a) uncle-niece marriages; (b) first cousin marriages (parallel and cross-cousin marriages); and (c) consanguineous marriages more distant than first cousins, *i.e.*, second cousins and beyond.

The collected data were then coded and entered in an IBM PC for analysis using the standard SPSS/PC$^+$ programme package (Statistical Package for Social Sciences). The data were analysed for incidence and type of consanguinity, coefficient of inbreeding, effects of consanguinity on fertility and child survival.

The degree of consanguinity, measured by the coefficient of inbreeding (F) is defined as the probability that a pair of genes at any particular locus in an offspring is derived from a single gene present in one of the ancestors common to both parents (Sanghvi, 1966). The F values of the offspring from marriages between uncles and nieces, between first cousins and between second cousins are 0.1250, 0.0625 and 0.0156 respectively. The weighted average of all inbreeding coefficients of the progeny including those with F = 0, provides the mean inbreeding coefficient ($\propto$) of a population, the weights (p_i) being the relative frequency of the progeny with inbreeding coefficient (F_i). Thus,

$$\propto = \Sigma p_i F_i$$

In practice, ∝ is estimated from the marriages rather than from the progeny (Roychoudhury, 1976b). The patterns of consanguinity and mean inbreeding coefficients that have been observed in study populations are presented in Table 14.1.

Results and Discussion

In the Ladakh region of Jammu and Kashmir state, consanguineous marriages seemed to be prescribed among Buddhist Bodhs; but practised, though not widely, amongst various muslim population groups. The frequencies of consanguineous marriages were found varying from a low of 14.8 per cent among Brokpas, to a high to 21.8 per cent among Arghuns. The mean inbreeding coefficients were found 0.0075, 0.0111and 0.0140 in Brokpas, Baltis, Arghuns respectively (Table 14.1). The most preferred type of consanguineous marriages amongst all was that between first cousins; more between cross-cousins than between parallel cousins. A few Balti, Brokpa, Arghum women (respondents) have also reported marriages between second cousins. Uncle-niece marriages have also been found in case of a few Arghuns (3 per cent) and Baltis (1 per cent), although this type is forbidden in Islam, as also reported by Roychoudhry (1976a) among Muslims of Jammu and Kashmir state. The Muslims (Pooled) group registered that the percentage of women marrying a relative was not exactly high (19.3 per cent); as the overwhelming majority of marriages were of non-consanguineous type (Table 14.1). The mean inbreeding coefficient for the Pooled data was estimated as 0.0116.

Table 14.1: Incidence of consanguineous marriages[1] and coefficients of inbreeding among Muslim population groups of Ladakh region in Jammu and Kashmir

Population	*Non-consanguinous-marriage*	*Consanguinous marriage*	*Type of consanguineous marriage*	*First cousin*			*Mean Inbreeding co-efficient*
			Uncle-niece	*Parallel cousin*	*Cross-cousin*	*Second cousin*	
Balti	81.2	18.8	0.7	7.0	8.7	2.4	0.0111
Brokpa	85.2	14.8	-	3.7	7.4	3.7	0.0075
Arghun	78.1	21.8	2.6	4.0	12.6	2.6	0.0140
Muslim (Poded)	80.7	19.3	1.2	5.8	3.7	2.6	0.0116

1: In percentages.

Table 14.2: Incidence of consanguinity and coefficients inbreeding among Muslims of India

Sl.No.	Country/Region/ State/Union Territories (U.T.)	Population group	Sample size	Consanguineous marriages (in %)	Coefficients of inbreeding	Authors
North India						
1.	Jammu and Kashmir	Balti	298	18.8	0.0111	Present study
		Brokpa	54	14.8	0.0075	Present study
		Arghun	151	21.8	0.0140	Present study
		Ladakh Muslim	503	19.3	0.0116	Present study
		Kashmiri Muslim	46	23.9	0.0149	Bhasin and Nag, 1999
		Gujjar	29	13.8	0.0086	Bhasin and Nag, 1999
		Ahmediyya	300	38.7		Kashyap, 1976
		Muslim	3230	18.6	0.012	Roychoudhury, 1976b
2.	Himachal Pradesh					
3.	Punjab	Muslim	327	1.22	0.0010	Roychoudhury, 1976b
4.	Chandigarh					
5.	Haryana					
6.	Delhi (U.T.)	Urban Sunni Muslim (Post-partition)	140	37.9		Basu and Ray, 1972
		Urban Sunni Muslim (Pre-partition)	572	15.9		Basu and Ray, 1972
		Sheikh Sunnis (Urban)	1158	29.1	0.0124	Basu, 1975
		Moghul Sunnis (Urban)	253	22.1	0.0112	Basu, 1975
		Pathan Sunnis (Urban)	72	23.6	0.0112	Basu, 1975

contd

table 14.2 contd.

Sl.No.	Country/Region/ State/Union Territories (U.T.)	Population group	Sample size	Consanguineous marriages (in %)	Coefficients of inbreeding	Authors
North India						
7.	Uttar Pradesh	Sayyad Shia (Urban)	1000	49.4	0.0199	Basu, 1975
		Muslim	1231	5.9	0.004	Roychoudhury, 1976b
		Navadic Gujjars (Postonelist)	112	61.6	0.037	Bandyopadhyay *et al*.. 1986
8.	Rajasthan	Bohra Muslim	412	41.3	0.022	Basu, 1976
West India						
9.	Gujarat	Muslim	753	39.7	0.025	Roychoudhury, 1976b
10.	Maharashtra	Momin	198	11.1	0.0066	Malhotra *et al*., 1977
		Memon	84	23.8	0.0159	Malhotra *et al*., 1977
		Khoja	95	14.7	0.0098	Malhotra *et al*., 1977
		Bohra	285	19.6	0.0124	Malhotra *et al*., 1977
		Irani	57	24.6	0.0175	Malhotra *et al*., 1977
		All Muslim	719	17.5	0.0094	Malhotra *et al*., 1977
		Memans	443	27.1	0.0122	Sanghvi *et al*., 1956
		Bohra	493	26.0	0.0126	Sanghvi *et al*., 1956
		Khoja	500	13.0	0.0064	Sanghvi *et al*., 1956
		Muslim and Parsi	2014	20.7	0.010	Sanghvi *et al*., 1956
11.	Goa, Daman and Diu					
12.	Dadra and Nagar Haveli (U.T.)					

table 14.2 contd.

Sl.No.	Country/Region/ State/Union Territories (U.T.)	Population group	Sample size	Consanguineous marriages (in %)	Coefficients of inbreeding	Authors
East India						
13.	Arunachal Pradesh					
14.	Assam	Bengali Muslim	203	3.0	0.003	Mukherjee & Chakravarty, 1977
15.	Nagaland					
16.	Manipur					
17.	Mizoram					
18.	Tripura	Muslim	693	5.2	0.003	Roychoudhury, 1976b
19.	Meghalaya					
20.	Sikkim					
21.	West Bengal	Muslim	471	19.3	0.0067	Barua, 1976
		Morriabad Muslim	145	26.9	0.0158	Huq, 1976
		Megha Muslim	187	25.1	0.0147	Huq, 1976
		Patket-Danga Muslim	175	13.1	0.0074	Huq, 1976
		Kabiraj Pur Muslim	131	17.6	0.0106	Huq, 1976
		Ekdala Muslim	118	28.0	0.0166	Huq, 1976
		Lalita Kundu Muslim	79	25.3	0.146	Huq, 1976
		West Bengal Muslim	835	22.2	0.135	Huq, 1976
		Muslim	63	5.9	0.004	Roychoudhury, 1976b

contd.

table 14.2 contd.

Sl.No.	Country/Region/ State/Union Territories (U.T.)	Population group	Sample size	Consanguineous marriages (in %)	Coefficients of inbreeding	Authors
East India						
22.	Bihar	Muslim		55.0	0.027	Anjani and Roy, 1989
		Muslim	388	9.5	0.006	Roychoudhury, 1976b
22.	Orissa					
Central India						
24.	Madhya Pradesh	Muslim and Bohra	351	59.3	0.0263	Goswami, 1970
		Muslim	173	60.1	0.0287	Goswami, 1970
		Bohra Muslim	178	52.8	0.0241	Goswami, 1970
South India						
25.	Karnataka	Muslim (Hospital Patients)	616	22.4		Devi *et al.*, 1981
		Muslim (Hospital Patients)	642	22.0	0.0131	Devi *et al.*, 1981
		Muslim	76	27.6	0.018	Roychoudhury, 1976b
		Muslim		25.6	0.0159	Bittles *et al.*, 1985
		Muslim	10789	27.7	0.0173	Bittles *et al.*, 1987
		Muslim	43	46.5		Devi *et al.*, 1987
26.	Andhra Pradesh	Shaikh and Syed	77	24.7	0.0154	Mukherjee *et al.*, 1977
		Muslim	7	28.6	0.018	Mukherjee *et al.*, 1977
		Urban Muslim		24.7	0.015	Mukherjee *et al.*, 1977
		Muslim	40	47.5	0.0296	Dube, 1960
		Muslim	5	20.0		Dromamraju & Meera Khan, 1960
		Muslim	65	46.2	0.030	Roychoudhury, 1976b
		Muslim	356	36.2	0.025	Sanghvi, 1966

table 14.2 contd.

Sl.No.	Country/Region/ State/Union Territories (U.T.)	Population group	Sample size	Consanguineous marriages (in %)	Coefficients to inbreeding	Authors
27.	Tamil Nadu	Muslim	983	34.2	0.021	Roychoudhury, 1976b
		Urban Muslim	549	17.5		Ramesh *et al.*, 1989
		Vellore Muslim			0.0019	Asha Bai *et al.*, 1981
		Rural Muslim (N. Arcot)			0.0201	Rao, 1983
		Urban Muslim (N. Arcot)			0.0113	Rao, 1983
		Madras Muslim (Moplah)	500	24.0		Chakravarthi, 1968
		Rural Muslim (N. Arcot)	212		0.0201	Rao and Inbaraj, 1977
		Urban Muslim (N. Arcot)	1960		0.0123	Rao and Inbaraj, 1977
		Muslim (N. Arcot)	6116	19.6		Rao *et al.*, 1971
		Muslim (Vellore town)	15	60.0		John and Jayabal, 1971
		Muslim	983	34.2	0.021	Roychoudhury, 1976b
		Rural Muslim (N. Arcot)	120	37.7		Rao *et al.*, 1972
		Urban Muslim (N. Arcot)	1445	19.4		Rao *et al.*, 1972
28.	Kerala	Muslim	215	21.4		Ali, 1958
		Malayalam Muslim	45	26.7	0.0167	Ali, 1958
		Tamil Muslim	35	25.7	0.0130	Ali, 1958
		Urdu Muslim	42	31.0	0.141	Ali, 1958
		Muslim	562	17.3	0.011	Roychoudhury, 1976b
29.	Pondicherry (U.T.)					
Islands						
30.	Lakshdweep (U.T.)	Koya Muslim	176	33.5	0.021	Roychodhury, 1977
		Malwi Muslim	435	4.1	0.002	Roychodhury, 1977
		Melachari Muslim	1471	6.3	0.004	Roychodhury, 1977
31.	Andaman and Nicobar Islands (U.T.)					

The main reasons behind this practice, which have been noted are: rural set-up, maintenance of social and economic status by upper classes, low socio-economic status of the lower classes; strengthening of ties between families, social solidarity; traditionalism, conformity to Islamic customs; maintenance of family property and land; parental domination in decision-making and arranging marriages; extended family structure; preference for mates within kin groups; low level of educational attainment. Similar reasons for the practice of consanguinity have been cited by Basu and Roy (1972), Basu (1975), in their study among Delhi Muslims; by Huq (1976) among Muslims of West Bengal. Several other have also endorsed similar viewpoints (Dronamraju and Meera Khan, 1961, 1963; Centerwall *et al.*, 1969; Rao *et al.*, 1972; Rao and Inbaraj, 1977; Devi *et al.*, 1982; Pingle, 1983; Reddy, 1983; Rao and Murthy, 1984; among others). However, it has been reported that there seems a general tendency of decline in the incidence in recent years, due to such factors as, spread of education; increasing mobility and communication facilities; as well as deviation from the traditional ways of mate selection etc. although specific study on the temporal changes is needed for a proper assessment.

It may be mentioned here that, in the Jammu and Kashmir state, whereas the frequency of consanguineous marriages among Ahmediyya Muslims (38.7 per cent) reported by Kashyap (1976) seems rather high; the frequency among Muslims in general, has been found only 18.6 per cent by Roychoudhury (1976a), which appears not much different from the frequency of consanguineous marriages in Ladakh Muslims as a whole. These studies have also mentioned that the most frequent type of consanguineous marriage is between first cousins, similar to the finding of the present study.

The comparison of the frequencies of consanguineous marriages across the northern region of India (Table 14.2) has revealed that in the nomadic Muslim Gujjars of Garhwal Himalayas in Uttar Pradesh there exist relatively very high frequency of consanguinity (61.6 per cent) (Bandyopadhyay *et al.*, 1986). And despite physical isolation, and other socio-economic constraints, the Muslims of Ladakh region are practising the same with much lesser frequency (19.3 per cent) than the Kashmiri Muslims (23.9 per cent - Bhasin and Nag, 1999)and Ahmediyya Muslims of Jammu and Kashmir (38.7 per cent - Kashyap, 1976); as well as muslims of Delhi (range - 22.1 to 37.9 per cent - Basu and Roy, 1972; Basu, 1975;

Krishan, 1975); Uttar Pradesh (49.4 per cent - Basu, 1975; 61.6 per cent - Bandyopadhya *et al.*, 1986); Rajasthan (43.0 per cent - Roychoudhuri, 1976).These data additionally indicate that such marriage practice is not confirmed to isolates and geographically restricted communities but is preferred even in numerically large groups without similar constraints.

Table 14.2 also presents an overall view of practice of consanguinity, among Muslims across various regions/states of India. And, it may be noted that the Gujjar Muslims of Uttar Pradesh still register the highest proportion of such marriages, whereas the Bengali Muslims of Assam have shown the lowest proportion (3 per cent - Mukherjee and Chakravorty, 1977).

Although no clear geographical trend in consanguinity is discernible, Table 2 broadly projects eastern region of India as a zone of low to moderate practice of consanguinity. All other region portrays moderate to high incidence of such marriages.

Further, studies on the effects of consanguineous marriage on fertility have also shown varying findings. Darlington (1960) postulated that human stocks could maintain the highest fertility with regular consanguineous marriages. And, elevated fertility levels in such marriages than in non- consanguineous ones have been observed in various populations of the world by many authors (Eaton and Mayer, 1954; Book, 1956; Schull *et al.*, 1970; Freie-Maia and Azevedo, 1971; Hussein, 1971; Shcull and Neel, 1972; Phillippe, 1974; Shami *et al.*, 1990).

In India, higher fertility levels in consanguineous marriage were noticed in Delhi and Uttar Pradesh by Basu (1975); in Orissa by Reddy (1979); in Karnataka by Devi *et al.*, (1981), Saheb *et al.*, (1981), Bittles *et al.* (1985, 1987), Hann (1985); in Andhra Pradesh by Reddy and Rao (1978), Ray (1979), Reddy and Reddy (1979), Sirajuddin and Basu (1984). Srikumari *et al.* (1985), (Reddy (1987); in Tamil Nadu by John and Jayabal (1971), Asha Bai *et al.* (1981) (among others). Among the Ladakh Muslims as well, in the present study, fertility (number of children ever born) seems higher in consanguineous marriages than in non-consanguineous ones (Table 14.3).

Hence, the assumption that consanguinity is confined mostly to areas of low population density and leads to decreased fertility (Cavalli-Sforza and Bodmer, 1971) appear nullified. Recently, Rao

(1991) concluded that inbreeding has not resulted in a marked decrease in fertility in South India, except in pre-selected groups such as, sick children or among the mentally retarded individuals. And therefore, rather than disturbing the established social practice of consanguinity (in south India) it would be advisable to provide genetic counselling to those families that have an affected child.

Table 14.3: Average number of children ever born[1] and surviving[1] and proportion of surviving children[1] among Muslims of Ladakh region in Jammu and Kashmir

Sl.No.	Demographic Parameter	Type of Marriage	Muslim population group			
			Baltis	Biskys	Arghuns	(Pooled)
1.	Average number of children born[1]	Non-consanguineous	4.50	4.98	3.57	4.28
		Consanguineous	4.91	5.13	4.18	4.68
2.	Average number of children surviving[1]	Non-consanguineous	3.21	3.46	3.03	3.19
		Consanguineous	3.29	2.63	3.21	3.21
3.	Proportion of surviving children	Non-consanguineous	0.713	0.695	0.849	0.745
		Consanguineous	0.670	0.513	0.768	0.686

1 = Estimate per married woman (respendent)

However, it may be noted that lower fertility levels in consanguineous marriages than in non-consanguineous ones have also been reported in many studies (Bemiss, 1858; Marcallo *et al.*, 1964; Post, 1965; Centerwall and Centerwall, 1966; Kumar *et al.*, 1967; Conterio, 1969; Ghosh, 1972; Barua, 1976; Reid, 1976; Ansari and Sinha, 1978; Devi *et al.*, 1981; Srikumari *et al.*, 1985).

Similarly, the studies on consanguinity and child mortality and survival also have been found varying. Many studies from across the world have reported elevated mortality rates in the offspring of consanguineous marriages as compared those of non-consanguineous ones (Bemiss, 1858; Sutter and Tabah, 1953; Schull *et al.*, 1970; Hussein, 1971; Seemanova, 1971; Shami *et al.*, 1990; among others). In India too, a host of studies have observed the same trend (Stevenson *et al.*, 1966; Kumar *et al.*, 1967; Chakravartti *et al.*, 1971; Basu, 1975, 1978; Reddy, 1979; Saheb *et al.*, 1981; Sirajuddin and Basu, 1984; among others. But on the other hand, some studies have also reported that parental consanguinity does not contribute to increased offspring mortality or decreased proportion of survivors (Cook and Hanslip, 1966; John and Jayabal, 1971; Bittles *et al.*, 1985). In the present study too, the proportion of surviving children appears lower in consanguineous than in non-consanguineous marriages, as seen in many other regions (Table 14.3).

Thus, the above results and discussion points that the incidence of consanguinity is relatively low in Ladakh Muslims despite physical isolation and various socio-economic constraints, and probably several factors are involved in the mate - selection process. Further, even though increased fertility and decreased proportion of surviving children have been observed in consanguineous than in non-consanguineous marriages among Ladakh Muslims, as also noticed elsewhere, a variety of other independent determinants have been found to be influencing these as well (Bhasin and Nag, 1999).

References

Ahmed, A.H. (1979). Consanguinity and schizophrenia in Sudan. Brit. J. Psychiatry, 134: 635.

Al-Awadi, S. A., Naguib, K. K., Moussa, M.A., Farag, T.I., Teebi, A.S. and Fl-Khalifa, M.Y. (1986). The effect of consanguineous marriages on reproductive wastage. Clin. Genet., 29: 384.

Alfi, O.S., Chang, R. and Azen, S.P. (1980). Evidence for genetic control of non-disjunction in man. Am. J. Hum. Genet., 32: 477.

Ali, S.G.M. (1968). Inbreeding and endogamy in Kerala (India). Acta. Genet. Stat. Med., 18: 369.

Anjani, A.V.S.S.M. and Roy, K. (1989). A study of genetical demographic parameters in the Muslims of Parastoli, Ranchi. Abstract presented at 14th Indian Society of Human Genetics Conference, Ranchi, 1989.

Ansari, N.A. and Sinha, S.P. (1978). Survey on the effects of inbreeding in two populations of Bihar. Ind. J. Med. Res., 68: 295.

Amer, G.B.L. Consanguineous marriages in the American population. Columbia University Studies in History, Economics and Public Law, 31:1.

Asha Bai, P.V., John, T.J. and Subramaniam, V.R. (1981). Reproductive wastage and developmental disorders in relation to consanguinity in South India. Trop. Geog. Med., 33: 275.

Bailit, H.L., Damon, S.T. and Amon, A. (1966). Consanguinity on Tristan da Cunha in 1938. Eugen. Quart., 13: 30.

Bandyopadhyay, S.S., Chakravarty, S.K.and Lakshmit, G.R. (1986). Consanguinity among pastoral nomadic Gujjars of north-westhern Uttar Pradesh. Hum. Sci., 35: 134.

Barua, S. (1976). Selection intensity among the consanguineous and non-consanguineous groups of a Muslim population of 24 Parganas, West Bengal. Man in India, 56: 359.

Basu, S.K. (1975). Effects of consanguinity among the North Indian Muslims. J. Pop. Res., 2 : 57.

Basu, S.K. (1978). Effects of consanguinity among Muslim groups of India. pp. 173. In : Medical Genetics in India. Volume 2. I.C. Verma (Ed.). Auroma Enterprises, Pondicherry.

Basu, S.K. and Roy, S. (1972). Change in the frequency of consanguineous marriages among the Delhi Muslims after partition. East. Anthrop., 25: 21.

Bell, J. (1940). A determination of the consanguinity rate in the general hospital population of England and Wales. Ann. Eugen., 10: 370.

Bemiss, S.M. (1858). Report on influence of marriages on consanguinity upon offspring. Trans. Am. Med. Assoc., 11: 319.

Bhasin, M.K. and Nag, Shampa. (1999). Anthropo-Demographic study among the population groups of Ladakh region in Jammu and Kashmir, India. 2. Fertility differentials and determinants. J. Hum. Ecol. (In Press).

Bhasin, M.K. and Nag, Shampa. (1999). Anthropo-Demographic study among the population groups of Ladakh region in Jammu and Kashmir, India. 3. Mortality differentials and determinants. J. Hum. Ecol. (In Press).

Bittles, A.H., Devi, A.R.R., Savithri, H.S., Rajeshwari, S. and Rao, N.A. (1985). Inbreeding and post-natal mortality in South India: Effects on the gene pool. J. Genet., 64: 135.

Bittles, A.H., Devi, A.R.R., Savithri, H.S., Sridhar, R. and Rao, N.A. (1987). Consanguineous marriage and postnatal mortality in Karnataka, South India. Man, 22: 736.

Bonne, B. (1963). The Samaritans - A demographic study. Hum. Biol., 35: 61.

Book. J.A. (1956). Genetical investigations in a north Swedish population: The offspring of first cousin marriages. Ann. Eugen., 21: 191.

Centerwall, W.R. and Centerwall, S.A. (1966). Consanguinity and congenital anomalies in South India: A pilot study. Ind. J. Med. Res., 54: 1160.

Centerwall, W.R., Savarinathan, G., Mohan, L.R., Booshanan, V. and Zachariah, M. (1969). Inbreeding patterns in rural South India. Soc. Biol., 16: 81.

Chakravartti, M.R. (1968). Consanguinity in India. Z. Morph. Anthrop., 60: 170.

Chakravartii, M.R., Rao, N.V.K., Satyanarayana, M. and Veeraju, P. (1971). A note on inbreeding in three villages on coastal Andhra Pradesh. East. Anthrop., 24: 71.

Conterio, F. (1969). Some results on fertility of consanguineous marriages. Atl. Ass. Genet. Ital. 14: 90.

Cook, R. and Hanslip, A. (1966). Mortality among offspring of consanguineous marriage in rural areas of East Jordan. J. Trop. Pediat., 11 : 95.

Darlington, C.D. (1960). Cousin marriage and the evolution of breeding systems in man. Heredity, 16: 297.

Darwin, G.H. (1875). Marriage of first cousins in England and their effects. J. Stat. Soc. Lond., 38 : 153.

Devi, A.R.R., Rao, N.A. and Bittles, A.H. (1981). Consanguinity, fecundity and post-natal mortality in Karnataka, South India. Ann. Hum. Biol., 8: 469.

Devi, A.R.R., Rao, N.A. and Bittles, A.H. (1982). Inbreeding in the State of Karnataka, South India. Hum. Hered., 32: 8.

Devi, A.R.R., Rao, N.A. and Bittles, A.H. (1987). Inbreeding and incidence of childhood genetic disorders in Karnataka, South India. J. Med. Genet., 24: 362.

Dolinar, Z. (1960). Micro-evolution among the Susak islanders. Inbreeding, sterility, blood groups and red hair. Ann. Hum. Genet., 24: 15.

Dronamraju, K.R. and Meera Khan, P. (1960). Inbreeding in Andhra Pradesh. J. Hered., 51: 239.

Dronamraju, K.R. and Meera Khan, P. (1963). Inbreeding in Andhra Pradesh (1961). pp. 126. Proc. 2nd Int. Cong. Hum. Genet.

Dronamraju, K.R. and Meera Khan, P. (1963). A Study of Andhra marriages - consanguinity, caste illiteracy and bridal age. Acta. Genet. State. Med., 13: 21 (1963).

Eaton, J. and Mayer, A. (1954). Man's capacity to reproduce. Hum. Biol., 25: 1.

Freire-Maria, N. and Azevedo, J.B.C. (1971). The inbreeding load in Brazilian Whites and Negroes as estimated with sib and cousin controls. Am. J. Hum. Genet., 23: 1.

Ghosh, A.K. (1972). Inbreeding in the Kota of Nilgiri Hills, Madras. Soc. Biol., 19: 289.

Ginter, E.K., Garkavtzeva, R.F. and Rarezov, A.A. (1980). Population structure and hereditary diseases in Uzbekistan. pp. 327. In: Population Structure and Genetic Disorders. A.W. Eriksson, A.R. Forsius, H.R. Nevalinna, P.L. Workman and, R.K. Norio (Eds.). Academic Press, New York.

Glass. B. (1954). Genetic changes in human populations especially those due to gene flow and genetic drift. Adv. Genet., 6: 95.

Goswami, H.K. (1970). Frequency of consanguineous marriages in Madhya Pradesh. Acta. Genet. Med. Gem., 19: 486.

Habib, Z. and Book, J.A. (1983). Consanguinity and incidence of thalassemia in Egypt. Hereditas. 99: 215.

Hafez, M., El-Tahan, A., Awadalla, M., El-Khayat, H., Abdel-Gafar, A. and Ghoneim, M. (1983). Consanguineous marriages in the Egyptian population, J. Med. Genet., 20: 58.

Hann, K.L. (1985). Inbreeding and fertility in a South Indian population. Ann. Hum. Biol., 12: 267.

Huq, F. (1976). Consanguinity and inbreeding among the Muslims of Murshidabad and Birbhum districts of West Bengal. J. Ind. Anthrop. Soc., 11: 21.

Hussein, F.H. (1971). Endogamy in Egyptian Nubia. J. Biosoc. Sci., 3: 251.

Ishikuni, N., Nemoto, H., Neel, J.V., Drew, A.L., Yanase, T. and Matsumoto, Y.S. (1960). Hosojima. Am. J. Hum. Genet., 12: 67.

Jonh, J.T. and Jayabal, P. (1971). Foetal and child loss in relation of consanguinity in southern India. Ind. J. Med. Res., 59: 1050.

Kashyap, L.K. (1976). A note on consanguinity among the Ahmediyyas of Kashmir. Res. Proc., 4: 107.

Krishan, G. (1975). Effect of parental consanguinity on anthropometric measurements among the Sunni Muslims of Delhi. Rec. Proc., 3: 4.

Kumar, S.K., Pai, R.A. and Swaminathan, M.S. Consanguineous marriage and genetic load due to lethal genes in Kerala. Ann. Hum. Genet., 52: 141.

Mahapatra, G.B. (1966). Consanguinity: A review. Ind. J. Paediat., 53: 275.

Malhotra, K.C., Ahamadi, K., Kazi, R.B. and Bhosale, N. (1977). Consanguineous marriages among the Muslim isolates of Maharashtra. J. Ind. Anthrop. Soc., 12: 207.

Marcallo, F.A., Freire-Maria, N., Azevedo, J.B.C. and Simoes, I.A. (1964). Inbreeding effect on mortality and morbidity in South Brazilian populations. Ann. Hum. Genet., 27: 203.

Mukherjee, D.P., Bhaskar, S. and Lakshmanudu, M. Studies in inbreeding and its effects on some endogamous populations of Chittoor district, Andhra Pradesh. Bull. Anth. Surv. India, 26: 10.

Naderi, S. (1979). Congenital abnormalities in newborns of consanguineous and non-consanguineous parents. Obstet. Gynaecol., 53: 195.

Neel, J.V., Kondavi, M., Brewer, R. and Anderson, R.C. (1949). The incidence of consanguineous matings in Japan with remarks on the estimation of comparative gene frequencies and the expected rate of appearance of induced recessive mutations. Am. J. Hum. Genet., 1: 156.

Pearson, K. (1902). Assortative mating in man. Biometrika, 2: 481.

Phillippe, P. (1974). Amenorrhoea, intra-uterine mortality and parental consanguinity in an isolated French-Canadian population. Hum. Biol., 46: 405.

Pingle, U. (1983). Comparative analysis of mating systems and marriage distance patterns in five tribal groups of Andhra Pradesh, India. Soc. Biol., 30: 67.

Post, R.H. (1965). Summary of workshop conference between demographs and population geneticists, under the auspices of American Eugenics Society, October, 1954. Eug. Quart., 12.

Ramesh, A., Sri Kumari, C.R. and Sukumar, S. (1989). Parallel cousin marriages in Madras, Tamil Nadu. New trends in Dravidian kinship. Soc. Biol., 36: 284.

Rao, A.P. and Reddy, V. (1983). Inbreeding among three endogamous groups in a multiscale village of Andhra Pradesh, India. Soc. Biol., 30: 109.

Rao, P.S.S. and Inbaraj, S.G. (1977). Inbreeding of Tamıı Nadu, South India, Soc. Biol., 24: 281.

Rao, P.S.S. and Inbaraj, S.G. (1977). Inbreeding effects on human reproduction in Tamil Nadu of South India. Ann. Hum. Genet. 41: 87

Rao, P.S.S. and Inbaraj, S.G. and Jesudian, G. (1972). Rural-urban differentials in consanguinity. J. Med. Genet., 9: 174.

Rao, P.S.S., Inbaraj, S.G. and Kaliaperumal, V.G. (1971). An epidemiological study of consanguinity in a large South Indian town. Ind. J. Med. Res., 59: 294.

Rao, V.V. and Murty, J.S. (1984). Selection intensities and inbreeding among some caste groups of Andhra Pradesh, India Soc. Biol., 31: 114.

Ray, A.K. (1979). Nature, amount and extent of consanguinity among two South Indian castes. J. Hered., 70: 281.

Reddy, B.M. (1979). Marriage distance and inbreeding in Vadabalija: A marine fisherfolk of Puri. Man in India, 59: 120.

Reddy, P.C. (1983). Consanguinity and inbreeding effects of fertility, mortality and morbidity in the Malas of Chitoor district. Z. Morph. Anthrop., 74: 45.

Reddy, P.G. (1987). Effects of consanguineous marriages on fertility among three endogamous groups of Andhra Pradesh. Soc. Biol., 34: 68.

Reddy, V.R. and Rao, A.P. (1978). Effects of parental consanguinity on fertility, mortality and morbidity among the Pattusalis of Tirupati, South India. Hum. Hered., 28: 226.

Reid, R.M. (1973). Social structure and inbreeding in a South Indian caste. p. 92. In: Genetic Structure of Populations. N.E. Morton (Ed.). University of Hawaii, Honolulu.

Roberts, D.F. (1956). A demographic study of a Dinka village. Hum. Biol., 28: 323.

Roychoudhury, A.K. (1976). Inbreeding in Indian populations. Trans. Bose. Res. Inst.,39: 65.

Roychoudhury, A.K. (1976b). Incidence of inbreeding in different states of India. Demography India. 5: 108.

Roychoudhury, A.K. (1977). Inbreeding in Lakshadweep. Man in India, 57: 328.

Roychoudhury, A.K. Consanguineous marriage in Tamil Nadu. J. Ind. Anthrop. Soc., 15: 167.

Saheb, S.Y., Sirajuddin, S.M., Raju, C.M., Sastry, D.B. and Gulati, R.K. Consanguinity and its effects on fertility and mortality in Kodava populations of Kodagu district, Karnataka. Ind. Anthropologist, 11: 101.

Sanghvi, L.D. (1954). Genetic diversity in the people of western India. Eug. Quart., 1: 235.

Sanghvi, L.D. (1966). Inbreeding in rural area of Andhra Pradesh. Ind. J. Genet., 26: 351.

Sanghvi, L.D., Varde, D.S. and Master, H.R. (1956). Frequency of consanguineous marriages in twelve endogamous groups in Bombay. Acta. Genet. Stat. Med., 6: 41.

Schull, W.J., Furusho, T., Yamamoto, M., Nagano, H. and Komatsu, I. (1970). The effect of parental consanguinity and inbreeding in Hirado, Japan. IV. Fertility and reproductive compensation. Hum. Genet., 9: 294.

Schull, W.J. and Neel, J.V. (1972). The effects of parental consanguinity and inbreeding in Hirado. Japan. V. Summary and Interpretation. Am. J. Hum. Genet., 24: 425.

Seemanova, E. (1971). A study of children of incestuous matings. Hum. Hered., 21: 108.

Shami, S.A., Schmitt, L.H. and Bittles, A.H. (1989). Consanguinity related parental and post-natal mortality of the population of seven Pakistani Punjab cities. J. Med. Genet., 26: 267.

Shami, S.A., Schmitt, L.H. and Bittles, A.H. (1990). Consanguinity, spousal age at marriage and fertility in seven Pakistani Punjab cities. Ann. Hum. Biol., 17: 97.

Sirajuddin, S.M. and Basu, A. (1984). Population structure, consanguinity and its effect on fertility and mortality among Chenchu of Andhra Pradesh. Hum. Sci., 33: 179.

Sphuler, J.N. and Kluckhonn, C. (1953). Inbreeding coefficients of the Ramah Navaho population. Hum. Biol., 25: 295.

Srikumari, C.R., Rajani Kumari, J. and Venkateswara Rao, T. (1985). Variability of genetic load with changing socio-cultural environment. Hum. Hered., 35: 388.

Stevension, A.C. and Warnock, H.A (1958-59). Observations on the results of pregnancies in women resident in Belfast. 1. Data relating to all pregnancies ending in 1957. Ann. Hum. Genet., 23: 382.

Sutter, J. and Tabah, L. (1953). Structure de la mortalite' dans les families consanguines. Population, 8 : 511 (1953).

Vogel, F. and Motulsky, A.G. (1986). Human Genetics: Problems and Approaches. Springer Verlag. Berlin. Heidelberg. New York. Tokyo. 2nd Edition.

Wright, S. (1922). Coefficients of inbreeding and relationship. Am. Natural., 56: 330.

15

Rural Economy, Sanitation and Health: A Review and Management

Nutan R. Biradar
Department of Family Resource Management,
University of Agricultural Sciences,
Dharwad - 580005

Introduction

In the society, man constantly interacts with his environment which efffects him both physically and psychologically. Our environment is getting degraded day-by-day mainly due to mans ill-actions/ill-habits that pollute the environment. The major types of pollutants which pollute the environment are air, water, noise and food pollutants. Inspite of various efforts made by different agencies, the people often remain ignorant.

To be able to live in a clean enviornment breathing clean air, drinking clean water and with accent to adequate sanitation facilities is among the humblest of human needs.

The world sanitation means "The science of safeguarding health". According to the National sanitation foundation of the U.S.A. sanitation is way of life. It is the quality of living that is expressed in the clean house, farm, business, neighbourhood and community. It was found that sanitary conditions of rural homes in India are reported to be very poor and unsatisfactory under different income groups. Rural people in India are still living under severe unhygienic conditions. Sanitation is one thing which needs to be thought much more in rural areas, as there found to be the practice of using open grounds for defecation. There are no proper drainage systems. The water used for bath, cleaning vessel etc., is just left to open space near the house.

Sanitation is next to saintliness. It is synonymously used with hygiene. Many organisations differ in their concepts on sanitation. Some international organisations conceptualise sanitation in terms of access to sanitation facilities like sanitary latrines. Some others go still further and widen the scope of sanitation by including components of soild and liquid waste disposal and treatment in its ambit. There are organisations who argue that sanitation include the entire health field and also that both are complementary and hence should go hand in hand.

The WHO defines environmental sanitation is as "the control of all those factors in mans physical environment which exercise or may exercise a deleterious effect on his physical development, health and survival". UNICEF has identified seven major components in sanitation. They present it as a package of health elements. This package includes, safe handling and usage of drinking water, disposal of waste water, treatment and dispoal of solid wastes and garbage, safe disposal of Human excreta, Home sanitation and Food hygiene, personal Hygiene and environmental (community) sanitation.

Environmental degradation has three damaging effects one, it is harmful to human health, two, reduces economic productivity and three, it leads to loss of amenities which are non-renewable (World Bank Reports). It has been projected in UNEP reports that around 70 per cent of the world's population in developing countires is without safe drinking water and its consequences are visible on human health due to waterborne diseases.

Health and sanitation are the other essential needs of human being, which exist along with food, clothing and shelter. Knowledge regarding health and sanitary practices is required for well being of an individual. But the women folk of rural areas are unaware about importance of sanitation. Therefore, there is need to generate awareness among women through educational means. SWACH, an integrated saniation water and community Health project envisages to undertake community education programme in the area of health and sanitation.

Open air defection is the single most reason for drinking water contamination. In open water sources, the faceal particles enter mainly through rain, which in hand pumps and pipe water supplies, the contamination is mainly through broken pipe and poor hygiene of platform surrounding the tap pump. When the water in the pump is not in continuous flow, the water in the surrounding area, near joints, leaks, etc., would enter the pipes through backward syphoning, especially during rainy season. Activities like bathing, brushing, ablutions, cleaning children near the taps/hand pumps are those contributing to contamination of drinking water. Waste water around taps/open unprotected wells are also consative factors.

The water thus contaminated could contain organisms causing diseases like diarrhoea typhoid, polio and hepatitis. These diseases in turn leads to poor health and loss of efficiency. It is estimated that 1800 million persons are lost in India every year due to diarrhocal diseases.

Reviews

Koushik and Verma (1982) carried out study on perceived feasibility of introducing sanitary latrines improving rural environment. The results revealed that out of the total respondents, 55.0 per cent have perceived two pit water seal sanitary latrine as most feasible for installation after they have been exposed to the innovation through action programme, whereas 30.8 per cent have expressed it to be moderately feasible. However, few (14.2 per cent) of respondents expressed this model as less feasible (Table 15.1).

Poonam and Oberoi (1992) conducted experiment on existing state of sanitations by taking 100 rural households of village in Punjab comprising 25 each from landless, small, medium and large land holders. They found that Sanitary conditions of medium and large farm size families in terms of type of heating units, provision of disposal of waste water, water stagnation in and around the house, place for garbage disposal, cattle shed, place of easing out and presence of insects in the house were better or satisfactory as compared to landless and small farm size families. The sanitary conditions in terms of provision of smokeless *Chulhas,* laterins in their homes and presence of insects, however, were found to be similar to all types of households (Table 15.2).

Table 15.1: Distribution of the households according to sanitary conditions of the different socio-economic status

Category	*Latrine*	*Drain*			*No drain*	*Cattle shed*	
		Covered	*Uncovered*			*Within house*	*Separately*
			Kacha	*Pucca*			
Landless	-	-	29 (58)	9 (18)	12 (24)	6 (12)	2 (4)
Small	-	-	28 (56)	12 (24)	10 (20)	11 (22)	3 (6)
Medium	4 (8)	-	32 (64)	18 (36)	-	12 (24)	4 (8)
Large	6 (12)	-	27 (54)	23 (46)	-	16 (32)	4 (8)

Category	*Bathroom*		*Area for garhage*	*House-hold pests in house*	*Pest control measures adopted*	*House electri-fied*
	Covered	*Open*				
Landless	1 (2)	49 (98)	-	50 (100)	11 (22)	-
Small	2 (4)	48 (98)	-	50 (100)	14 (28)	-
Medium	4 (8)	46 (92)	1 (2)	50 (100)	28 (58)	4 (8)
Large	9 (18)	41 (82)	2 (4)	50 (100)	42 (84)	12 (24)

Table 15.2: Distribution of households according to the health conditions of the family

Category	*Common diseases*	*Chronic diseases*	*Infectious diseases*	*Deficiency diseases*
Landless Labourers				
Children	44 (88.0)	3 (6.0)	12 (24.0)	10 (20.0)
Adults	38 (78.0)	45 (90.0)	11 (22.0)	18 (36.0)
Old	8 (16.0)	12 (24.0)	16 (32.0)	10 (20.0)
Small farmers				
Children	39 (78.10)	1 (2.0)	7 (14.0)	3 (6.0)
Adults	28 (56.0)	1 (2.0)	-	2 (4.0)
Old	13 (26.0)	-	9 (18.0)	1 (2.0)
Medium farmers				
Children	30 (60.0)	-	-	7 (14.0)
Adults	11 (22.0)	2 (4.0)	6 (12.0)	6 (12.0)
Old	12 (24.0)	5 (10.0)	-	11 (22.0)
Large farmers				
Children	11 (22.0)	-	2 (4.0)	-
Adults	18 (36.0)	2 (4.0)	1 (2.0)	2 (4.0)
Old	8 (16.0)	6 (12.0)	2 (6.0)	-

Sidhu *et al.* (1999) studied the effect of sanitations on health status of rural households in Udaipur district of Rajasthan (Table 15.3). They concluded that the sanitation in rural areas is not satisfactory. Further, they suggested that Government should provide the rural areas with the proper sewerage system. Village panchayats will have to put more efforts and resouces to improve the overall sanitation of the villages. The extension workers should educate the rural residents about the ways and advantages of living in clean sanitary environment.

Table 15.3: Existing state of sanitation in the rural households

Sanitary aspect	*Landless*	*Small*	*Medium*	*Large*
Floor area/person (m^2)	5.37	8.56	9.26	15.40
Heating units				
(a) Traditional chulha	25.00 (100)	25.00 (100)	24.00 (96)	24.00 (96)
(b) Smokeless chulha	-	-	-	1 (4.00)
(c) Kerosene stove	1 (4.00)	2 (8.00)	4 (16.00)	2 (8.00)
(d) Gas	-	1 (4.00)	2 (8.00)	3 (12.00)
Smoke outlet	2 (8.00)	2 (8.00)	4 (16.00)	8 (32.00)
Disposal of waste water				
(a) Kutcha drains	14 (56.00)	13 (52.00)	18 (72.00)	15 (60.00)
(b) Pucca drains	-	5 (20.00)	4 (16.00)	8 (32.00)
Water stagnation	12 (48.00)	10 (40.00)	2 (8.00)	5 (20.00)
Garbage disposal				
(a) In the house	2 (8.00)	1 (4.00)	1 (4.00)	-
(b) In the street	3 (12.00)	2 (8.00	-	-
(c) In the common community area	20 (80.00)	22 (88.00)	24 (96.00)	-
Cattle shed				
(a) Within residental area	11 (44.00)	13 (52.00)	6 (24.00)	25 (100.00)
(b) Away from residential area	1 (4.00)	2 (8.00)	14 (56.00)	19 (76.00)
Easing facility				
(a) Latrine	1 (4.00)	3 (12.00)	6 (24.00)	8 (32.00)
(b) Open field	24 (96.00)	22 (88.00)	19 (76.00)	17 (68.00)
(c) Street side	11 (44.00)	8 (32.00)	5 (20.00)	6 (24.00)

Figures in parantheses indicate percentages

Virk *et al.* (1996) studied the impact of modernisation on sanitary conditions of rural and urban inhabitants of Ludhiana district in Punjab (Table 15.4). Their study revealed that urban households had sewage connection from municipality and disposal of garbage was dustbins whereas rural houses and sewage connection linked with drains, soakage pits and baskets, compost pits were used for garbage disposal. Attached bathroom and toilets with modern communities were observed in urban houses whereas in rural house separate covered bathrooms were available with no amenities. Approximatley 50 per cent houses in far away villages had no toilet facility and eased out in open fields. Further, they concluded that rural areas lag behind in adoption of modern toilet facilities. Greater the distance from urban areas, lesser is the availability of these facilities. There is a need to strengthen extension education programmes for women and other residents of far-away villages, so that they understand the importance of waste disposal and toilet facilities.

Table 15.4: Provision for disposal of sewage, garbage and measures for protection of health (per cent)

Sanitary conditions	*Rural (villages) n = 100*		*Urban (localities) n = 100*	
	Near by (50)	*Far away (50)*	*Ordinary (50)*	*Posh (50)*
Sewage system linked with drain	56	84	-	-
Soakge pit	44	16	-	-
Municipality system	-	-	100	100
Garbage disposal dustbin	2	2	96	98
Basket	98	98	4	2
Composite pit	100	100	-	-
Outside trolley	-	-	100	100
Cattle sheds adjoining to residental area	26	58	-	-
Away from residential area	48	36	-	-
Not available	26	6	-	-
Control of insects/pests Providing wire-mesh doors/windows	50	30	100	100
Enough of sunlight	100	100	100	100
Using repellants	48	18	92	96
Insecticides	60	70	4	6

Dhillon *et al.* (1990) studied that the awareness among the housewives regarding sources of environmental pollution. The results of the study indicated that the rural respondents mainly came across environmental problems as a result of fuel burning, dust produced through house. The ill-effects caused by air, water and noise pollution and food poisoning were reported by both rural and urban respondents [Table 15.5(a)].

Table 15.5(a): Mean ranks of awareness of ill-effects caused by air and water pollution in rural and urban areas

Air pollution			*Water pollution*		
Ill effect	*Rural*	*Urban*	*Ill effect*	*Rural*	*Urban*
Irritation to the eyes	4.17 (I)	3.24 (I)	Sore throat	3.25 (I)	3.98 (III)
Suffocation	5.48 (II)	5.65 (VII)	Typhoid	3.81 (II)	3.88 (II)
Headache	5.77 (III)	4.66 (II)	Dysentery	4.18 (III)	4.80 (IV)
Bronchitis	6.65 (IV)	5.53 (VI)	Cholera	4.30 (IV)	5.41 (V)
Asthma	6.85 (VI)	5.58 (VI)	Goitre formation	4.33 (V)	3.48 (I)
Nausea	7.34 (VIII)	5.16 (III)	Tonsils	6.49 (VI)	5.66 (VII)
			Hepatitis	7.69 (VIII)	5.54 (VI)
Reduced workability due to dizziness	8.18 (X)	6.62 (IX)			
Soiliness of clothes and the household articles	7.85 (IX)	7.38 (X)	Inbreeding of mosquitoes and files	7.12 (VII)	7.01 (VIII)
Destruction of whitewash/ exterior paint	6.66 (V)	5.34 (IV)	Destruction of field crops	8.22 (IX)	8.48 (IX)
Destruction of vegetation	7.12 (VII)	5.44 (V)	Destruction of fisheries	10.00 (X)	10.00 (X)
r (ranks) = 0.82 per cent			r (ranks) = 0.82 per cent		

* significant at p = 0.01

Dhillon *et al.* (1990) concluded a case study on environmental management in rural and urban habitats in Punjab. They carried out survey in five colonies of Ludhiana city and five adjoining village of Ludhiana district revealed that the lack of family finance was the root cause of household pollution, followed by other factors, viz., place of dwelling, defective drainage system, improper disposal of refuse and traditional style of living. The factors affecting neighbouring environmental pollution included lack of civic sense, lack of governmental efforts industrialisation in the near vicinity and faulty habits of the people. Both rural and urban respondents strongly expressed the utility of LPG/Gobar gas for household cooking to avoid smoke drugery alongwith use of dried fuel, inbuilt kitchen as well as wire mesh doors, proper water facilities and drainage system for the improvement of household environment.

Table 15.5(b): Mean ranks of awareness of ill-effects caused by noise pollution and unhygienic food in rural and urban areas

Noise poillution			*Unhygienic food*		
Ill effect	*Rural*	*Urban*	*Ill effect*	*Rural*	*Urban*
Irritation to the ears/hard of hearing	2.06 (I)	2.52 (I)	Stomachache	2.30 (I)	2.30 (I)
Headache	2.74 (II)	2.53 (II)	Food poisoning	2.80 (II)	2.93 (II)
Annoyance	4.86 (VI)	3.90 (IV)	Dysentery	3.70 (III)	4.00 (IV)
Distrubed sleep	4.46 (IV)	3.74 (III)	Nausea	4.01 (IV)	3.25 (VI)
Reduced working efficiency	4.16 (III)	4.34 (V)	Vomiting	4.33 (V)	4.21 (VI)
Rise in blood pressure	6.53 (VII)	6.49 (VII)	Cholera	5.34 (VI)	4.54 (VII)
Restlessness	4.58 (V)	4.97 (VI)	Gastric ulcers	6.00 (VII)	4.18 (V)
r (ranks) = 0.82 per cent			r (ranks) = 0.85 per cent		

* singnificant at p = 0.05 ** significant at p = 0.01

Singal (1993) studied the knowledge, attitude and symbolic adoption of social forestry for promoting rural sanitation. The findings thus point towards the need for creating awareness among people regarding the role of social forestry in promoting environmental sanitation.

Solanki and Gupta (1999) studied the gain in knowledge by rural women through awareness camp on Environmental sanitation and they concluded that, there was significant gain in overall and component wise knowledge of the participates. It implies that the

Table 15.6(a): Mean ranks of factors associated with insanitary household conditions and neighbouring environmental pollution in rural and urban habitats

Insanitary household conditions			*Neighbouring environmental pollination*		
Factor	*Rural*	*Urban*	*Factor*	*Rural*	*Urban*
Economic status of the family	1.88 (I)	3.54 (I)	Lack of civic sense	2.08 (I)	2.86 (II)
Traditional style of living	3.64 (V)	6.50 (II)	Faculty habits of the people (use of drungs, alcohal)	4.10 (II)	5.68 (IV)
Place of dwelling	4.85 (III)	3.61 (III)	Lack of environmental education (courses in school/college curriculum)	5.21 (IV)	5.60 (V)
Defective drainage system	4.86 (IV)	4.41 (V)	Lack of governmental efforts for improving the environment (laws, acts, rules, campaign)	5.24 (V)	3.76 (III)
Improper disposal/placement of garbage and refuse	5.24 (V)	3.56 (II)	Industrialization in the residential area	4.90 (III)	2.54 (I)
Noise of heavy household equipment/ forage cutting machine/running tap water/TV and Radio	5.69 (VIII)	3.85 (IV)	Population explosion	5.73 (VI)	4.69 (IV)
Population explosion	5.61 (VII)	5.48 (V)	Exploitation of natural resouces (cutting of forests and shady trees)	6.81 (VIII)	5.96 (VII)
Miscellaneous factors (No use of wire mesh doors, open toilet pits, improper installation of hand pumps, lack of ventilation)	5.29 (VI)	6.58 (VIII)	Misuse and mishandling of the resources	7.85 (VIII)	7.06 (VIII)
			Much dependence upon advance technology (more use of pesticides and other chemicals)	8.00 (IX)	8.05 (IX)

camp had helped the rural women to gain and enrich their knowledge regarding environmental sanitation.

Table 15.6(b): Mean ranks of sources of inspiration for environmental management in rural and urban habitats

Source of inspiration	*Rural*	*Urban*
Friends and relatives	2.66 (I)	2.48 (I)
Radio programmes	4.13 (II)	4.05 (II)
Newspapers	4.33 (III)	4.17 (III)
TV Programmes	4.26 (IV)	4.57 (V)
Magazines	6.28 (V)	5.77 (VI)
Municipality efforts	10.00 (VIII)	6.96 (VIII)
Voluntary Organizations	7.54 (VII)	4.26 (IV)
Governmental efforts (formulation of environmental laws, acts, etc., offering courses in schools/colleges, celebrations of World Environment day, etc.	6.58 (VI)	6.21 (VII)

Sawkar (1989) conduct study on Health and Sanitation technologies to improve rural environment under All India Co-ordinated Research Project on Home and Farm Management at Rural Home Science College, University of Agricultural Sciences, Dharwad. In order to improve the sanitation, it was planned to introduce the improved low cost latrines, soakage pits and the improved drainage system. Through the film shows, pampletes followed by lecture on advantages of construction of low cost latrines, soakage pits and drainage system awareness in maintaining hygienic conditions was created. The results of the study revealed that, the attitude of rural households towards construction of latrines was negative. Their opinion was that, since most of their houses do not have open space around, the construction of latrines is not good for health. Demonstrations on soakage pit was given using course sand, gravels, pebbles, brick pieces and small stones for filling the soak pit. Along with soakage pits, demonstrations on improved drainage system was also given. The response to adopt this technology was favourable.

Ambeker (1999) was recommended to both the State and Central Government and also to the local bodies like the Municipal Corporation, Zilla Parishad, Taluka Panchayat and Gram

Panchayats, that projects related to providing unpolluted wholesome drinking water, closed drainage system, latrines, clean street and roads, waste disposal and treatment plants, etc. must be take up as a top priority area.

Conclusion

Remidial measures to overcome the sanitation and health problems:

(a) Government should provide the rural areas with the proper sewerage for the improvement of household environment.

(b) Village Panchayat will have to put more efforts and resources to improve the overall sanitation of the villages by introducing sanitation technologies like construction of low cost latrines, soakage pits and proper drainage system.

(c) Smokeless chulha and biogas are the important measures to control smoke which is disease causing agent and economical way of using fuel, respectively.

(d) The extension workers should educate the rural residents about the way and advantages of living in clean sanitary environment.

(e) Awareness compaign about sanitation and health to the rural women is essential.

(f) There is a great need to provide environmental education in the form of developing curriculam including courses on environment in schools and colleges.

(g) People's participation, role of bureaucrats, role of people's representatives, role of non-government organisation could change the scenario and fulfil our vision towards a clean global environment.

(h) Government should make prime efforts to involve public in the programmes *viz.* Van Mohotsav, World Earth Day and World Environment Day celebrations to encourage people to keep environment free from pollution.

References

Ambekar, J. (1999). Water pollution and epidemics in India: Tassards sustainable approach. *J. Hum. Ecol.,* 10(1): 35-39.

Dhillon, M.K., Dhaliwal, G.S. and Harpreet Cheema (1990). Environmental management in rural and urban habitats: A Case study of Punjab. *Indian J. Ecol.,* 17(2): 151-158.

Dhillon, M.K., Harpreet Cheema and Dhaliwal, G.S. (1990). Sources of environmental pollution in rural and urban habitats awareness among housewives. *Indian J. Ecol.*, 17(1): 13-16.

Kaushik, D. and Verma, T. (1992). Perceived feasibility of introducing sanitary latrines improving rural environment. *Indian J. Ecol.,* 9(1): 126-127.

Poonam and Krishna Oberoi (1992). Existing state of sanitation in rural households of village in Punjab. *Indian J. Ecol.,* 19(1): 94-98.

Ramadevi, O.T. (1998). Sanitation, water quality management and people's participation. CARITAS India, *Quarterly,* 3: 20-26.

Sawkar, S. (1989). *Annual Report,* All India Co-ordinated Research Projects, Family Resource Management, University of Agricultrual Sciences, Dharward, p. 64.

Sidhu, M., Virk, M.K. and Singh, S. (1999). Effect of saniation on health status of rural households: A study of Udaipur district of Rajasthan. *Indian J. Ecol.,* 26(1): 90-93.

Singal, S. (1993). Study of knowledge, attitude and symbolic adoption of social forestry for promoting rural sanitation: An action research. *Indian J. Ecol.,* 20(1): 6-66.

Solanki, D. and Gupta, P. (1999). Gain in knowledge by rural women through awareness camp on environmental sanitation. *Rural India,* pp. 135-137.

Virk, M.K., Dhillon, M.K. and Dhaliwal, G.S. (1996). Impact of modernisation on sanitary conditions of rural urban inhabitants of Ludhiana district. *Indian J. Ecol.,* 23(11): 1-6.

16

Impact of Industrialisation on Tribal Groups of Jharkhand

B.N. Pandey
P.G. Department of Zoology,
Purnia College, Purnia - 854 301 (Bihar)

According to 1991 Census, tribals constitute 8.03 per cent of the total population. The main concentration of tribal population in Bihar is in the Chotanagpur and Santhal Pargana divisions, where 90 per cent of the tribals of the state are concentrated. In Bihar the total population of tribals is around 5,810,867 and out of this totality 93.77 per cent live in rural areas and the rest 6.23 per cent live in urban areas.

Jamshedpur belt is highly industrialised having a tribal population of 4,66,572 people (male - 236318 and female - 230254). The literacy rate among male population is 38.36 per cent while in female it is 13.84 per cent. Major occupation of these tribals is farming. Besides, they also work as daily labourers in gravel mining, crusher mills and other industrial establishments. The main crop of the area is paddy. In these areas there is no proper means of irrigation. According to local people the agricultural products fulfil their needs only for six months and for the rest six months, they have to depend on other sources for their livelihood. In tribal belt, there is no proper facility of drinking pure water. Though in some villages the Government has provided hand pumps and wells, yet in some rural areas the people have to depend on natural sources of drinking water. Summer brings an acute scarcity of drinking water. During summer, people have to cover a distance of 5 to 6 kilometers to fetch drinking water. In summer season when most of the small rivers, streams, wells, as well as hand pumps dry up, the villagers dig river beds to the depth of 5 to 6 feet for water. The water of such pit remains so dirty that it is not possible to drink it without filtering. Such stagnant water contains a large number of zooplanktons, phytoplanktons, bacteria, fungi and insects, etc. Such stagnant water becomes the breeding ground of mosquito and ultimately leads to malaria. Due to the use of such water, the popluation suffers from gastro-intestinal disorders. It has been reported that nearly 83 per cent of tribal population are suffering from gastro-intestinal disorders. Water-borne diseases (like - cholera, jaundice, amoebiasis) are quiate common in the area.

The Kharia (one of the endangered tribals of Chotanagpur) have no agricultural land. They depend on forest for their livelihood. Their only source of income is to sale forest firewood. They earn Rs. 25-30 per day by selling wood and the major part of it goes to wine shops. They are fond of deshi liquor (Mahua wine). They are victims of malnutrition and suffer from T.B. and malaria. Physically they are very weak and have poor health. Infant mortality among them is high due to the lack of proper care and malnutrition.

The existence of Asur tribal commuity is threatened due to Bauxite mining in the area. The land of the said tribal community is being turned into fallow land at a great speed and extent by the base holders in convenience with the senior officials. The fertile land is becoming unfit for agricultural purposes. Consequently, Bauxite mining is proving to be a curse instead of boon for the Asur and other tribes of the plateau region. According to a survey report (published in "The Indian Nation") out of a total population of 10,712 in Bihar, 9,067 are found in Gumla district which consists of 42 villages. According to the report, 99.29 per cent land holders are uprooted by mining. A recent two day seminar on Bauxite Mining and Forest recommended a study of the impact of bauxite mining on the health of tribals along with the flora and fauna. According to experts, Bihar has about 61.10 million tonnes of bauxite reserves and about 11 lakh tonnes of bauxite is mined every year. About 2,500 acres of land of tribal regions were under lease in Gumla district alone dominated by Asur tribes. The dust particles of bauxite spread all over the area and settle down on trees and houses as a result of which their colour becomes red. It has been reported that due to a polluted environment and water the people of the area suffer from weak health, asthma, T.B., respiratory troubles and abortions. Due to mining, the land of the area is becoming barren leading to lower or nil production. In lack of proper food, tribals of the area have to depend on forest products for nutrition.

Out of 30 types of tribes found in Bihar, Birhor is one of the minor tribes found mainly in the districts of Koderma, Dhanbad, Hazaribagh, Giridih, Chatra and Ranchi. They lead a vagabond life and live in jungles. Their number is declining day-by-day. Due to their decreasing number, the Birhors have come to be known as Vanishing Tribe. Due to extremely high death rate and low birth rate, the total number of Birhors is now 6700 only. Due to increasing pollution and deforestation they are facing a lot of troubles in their livelihood and settlement.

Pahariya is another primitive tribe, chiefly found in Santhal Pargana. Their number is declining day-by-day and their existence is in danger. According to 1951 Census the total population of Paharia in Santhal Pargana was 222174 but presently their total population of Paharia has reduced nearby 1 lakh. They are victims of a large number of health hazards such as T.B., goitre, malaria, cancer and liver troubles. Due to improper medical facilities, they depend on plant medicines for treatment of different ailments. Their another problem relates to drinking water. The main sources of drinking water are ponds, streams and rivers which are utilised both for drinking purposes and for bathing. Due to this, they suffer from a variety of waterborne as well as skin diseases. Nearly 83 per cent mothers are unable to provide milk to their children and this is one of the major reasons of high infant mortality rate.

A recent survey (conducted by the author) in eight villages around Jadugoda reveals that the inhabitants of villages near Uranium Corporation of India Limited (UCIL), are suffering from a large number of abnormalities, like musculo-skeletal abnormality, leukemia, thalassaemia, disturbances in menstrual cycle and increasing rate of sterility in male. Apart from these, there is a high rate of abortions in the population residing near UCIL. It has been reported by villagers that whenever water of Tailing Pond (where waste products of Uranium are dumped) is used in irrigation, the plants dry up, generally after breeding one or two fruits. Further, they also pointed out that when Tailing Pond dries up, children usually play there. According to them, these children become victim of a large number of abnormalities.

The Jharia coal field occupies a distinct place on the coal map of the world. It is spread over an area of 450 kms and mining is being carried out in this coal field for more than 100 years. Recently it has been reported that despite extensive measures taken on curative part, a good number of miners suffer from various ailments. The most common diseases reported are chest disease, gastro-intestinal and allergic disorders. A recent survey carried by staff of MAB Project, Department of Zoology, Purnia College (Purnia) has revealed prevalence of malaria, tuberculosis, leprosy, gastro-intestinal disorder, nutritional deficiency, sexually transmitted diseases. It is to be noted that deforestation is one of the major factors of malaria. No doubt, due to deforestation tribals of East Singhbhum are also victim of malnutrition and due to ecological imbalances they have become victim of a larger number of diseases. Chakulia is a block of East Singhbhum having many soap and rice factories. The waste products of these factories are polluting the ground water, owing to a lack of proper disposal arrangement.

They are becoming landless in their own home. The process of indiscriminate exploitation has set in creating ethnic, nutritional, cultural, socio-economic problems as well as health status including reproductive fitness. No doubt, the industrialisation and mining of these mine rich segment of Bihar has given a lot, including the arrangement of bread and butter and certainly diseases, deforestation and ecological imbalances as bonus to the inhabitants of these areas which ultimately threaten the health status. It is true that the economic condition of the area has improved due to industrialisation but owing to the shy nature of tribals of the area, they are not getting full advantage of it. In concluding remarks it can be said that due to industrialisation, urbanisation, mining, deforestation and immigration, the socio-cultural, food habit and living standards have changed a lot. But in return they are getting a lot of health hazards. The day is not far when these primitive tribes will disappear from the area. It is, therefore, highly imperative for geneticists, anthropologists, ecologists, evolutionists, social scientists and medical researchers to collect, analyse and massive survey of data on ecology, sociology, nutritional, health status as well as genetic factors to save these disappearing tribes. Not only that, Government should also provide better drinking water facilities, employment as well as educational facilities so that these tribal people may be saved from the menance of industrialisation, mining and urbanisation.

17

Preliminary Phytochemical Analysis of Medicinal Plants of South Chotanagpur used Against Dysentery

R.L. Ram
Department of Botany (Centre for Biotechnology),
Marwari College, Ranchi - 834 001, Bihar

Abstract

A preliminary phytochemical anaylsis of medicinal plants has been carried out which act against dysentery. In the first phase of chemical analysis 10 medicinal plants have been screened whcih are used in the treatment of dysentery. It is found that the plants contain a number of secondary metabolites and have the pharmacological actions. Phytochemical analysis confirms that alkaloids and saponin promptly act against the treatment of dysentery.

Introduction

Chotanagpur plateau of Bihar is a tribal dominated area and is considered the paradise of botanists and is well known for its diverse medicinal plants. As such, this area is rich in its flora, fauna, soil and climate. Villagers have a great knowledge of the medicinal plants, used in various ailments. A survey of hospitals in and around Ranchi has revealed the facts that thousands of patients come with a complain of dysentery. It is assumed that the higher ferrous, maganese and magnesium content in the water is one of the major causes of gastrointestinal disorder in the form of dysentery. Such studies have recently been confirmed by many doctors. Children as well as adults also suffer from dysentery in this region. However, in the remote areas, people do not depend upon the Allopathic medicines rather they have effective herbal cure of the diseases without any risk or side effect. Mahato *et.al.* (1992) and Ram and Saha (1998) have worked on medicinal plants used in dysentery by the people of Chotanagpur. But extensive work on the medicinal plants used in dysentery have not yet been done by any researcher.

Therefore, an intestive survey with a view to identification, collection, systematic studies and chemical analysis of these plants would throw sufficient light on the role of different chemical constituents in the containment of the disease is needed, which indicates a growing interest among scientists to carry out ethnobotanical studies of this region.

Materials and Methods

The plants were collected during their flowering periods (1998-99) and identified with the help of standard flora.

Phytochemical screening approach of the plants has been done. The plants have been screened to show the presence or absence of phytochemical classes of compounds. The medicinal plants, used against dysentery, were screened for the presence of alkaloids, sterols, saponins, terpene and tannins by the conventional methods proposed by Bhattacharjee and Das, 1969.

Results and Discussion

In modern medicine system there are lots of use of herbal drugs for both permanent and temporary diseases. During the course of physiological activites, some chemicals in the form of secondary metabolites are synthesised in the plants. Most of the chemicals have a very significant therapeutic action on the human health system.

Keeping these things under consideration ten plants have been phytochemically screened. It was found that most of the plants contain a number of secondary metabolites and some of them have the pharmacological actions.

The preliminary phytochemical screening shows that the plant contains different types of chemical constitutents, having different therapeutic actions.

Rauwolfia serpentina is one of the plant in which a large number of alkaloids are present. Some of the alkaloids like reserpine and rescenamine is extensively used in modern medicine system as hypertensive drug. According to the vaidyas and kavirajas, root of *Rauwolfia serpentina* is also useful in dysentery.

From the preliminary chemical analysis Table 17.1 shows the presence or absence of the phytochemicals on preliminary chemical analysis which act in the treatment of dysentery. It has been found that tannin is present in the whole plant of *Aerva lanata* Juss; *Euphorbia hirta* Linn., root of *Rauwolfia serpentina* Benth. and leaves and root of *Trachyspermum ammi* (L) Sprague. In *Euphorbia hirta* Linn. and *Rauwolfia serpentina* Benth. alongwith tannin and alkaloids are also present.

Alkaloids are present in leaves of *Catharanthus roseus* G. Don, *Eclipta alba* Hassk, *Solanum nigrum* Linn., whole plant of *Elephantopus scaber* Linn., Bark of *Holarrhena antidysenterica* Wall.

Saponin is present in *Achyranthes aspera* L. This plant is used in various diseases but the leaves of *Achyranthes aspera* L. are useful in the treatment of dysentery.

In past many workers like Mukhopadhyay *et.al.* (1988), Eilert *et al.*, (1987), Smith *et al.*, (1987) and Benjamin *et al.*, (1990) have reported alkaloids from leaves and roots of *Catharanthus roseus* which confirms the present work.

Holarrhena antidysenterica is the most effective herbal medicine which act against dysentery by the use of stem bark that contains alkaloids. Many workers (Prasad and Kaul, 1957; Jai Shankar and Basu, 1961; Dutta and Iyer, 1968) have reported the presence of antiamoebic value of berberine and kurchi alkaloids that act against dysentery. Dhar and Bhattacharji (1955), Ruyter *et al.*, (1991) have investigated the presence of total alkaloid contents of *Rauwolfia serpentina* roots from regenerated plant seems to be slightly highest. Thakashi *et al.*, (1991, 1994) have studied the detailed chemical anaylsis in Euphorbiaceae. They found tannins and related polyphenols. Many workers (Lin-Geeng *et al.*, 1995, Yoshiaki and Yoshida 1996) have also confirmed the presence of hydrolyzable tannin, which also confirms the presence of tannin in *Euphorbia hirta*.

The present investigations are the obvious lines of future research to findout the active principles of biochemical analysis of medicinal plants, which act against dysentery.

Table 17.1: Preliminary Chemical Analysis of Medicinal plants used in the treatment of Dysentery

Botanical Name ** Local Name*	*Plant parts used*	*Steroid*	*Terpene*	*Alkaloid*	*Saponin*	*Tannin*
Achyranthes aspera Linn. * Chitchiti	Leaves	(-)	(-)	(-)	(+)	(-)
Aerva lanata Juss. * Kapurijari	Whole Plant	(-)	(-)	(-)	(-)	(+)
Catharanthus roseus G. Don. * Sadabahar G	Leaves, Root	(-)	(-)	(+)	(-)	(-)
Eclipta alba Hassk. * Bhringraj	Leaves	(-)	(-)	(+)	(-)	(-)
Elephantopus scaber Linn. * Jhorjhuti	Whole Plant	(-)	(-)	(+++)	(-)	(-)
Euphorbia hirta Linn. * Dudhi	Whole Plant	(-)	(-)	(+)	(-)	(+)
Holarrhena antidysenterica Wall * Kurchi	Bark	(-)	(-)	(+)	(-)	(-)
Rauwolfia serpentina Benth. * Sarpagandha	Root	(-)	(-)	(++)	(-)	(+)
Solanum nigrum Linn. * Makoy	Leaves	(-)	(-)	(+)	(-)	(-)
Trachys permum ammi (L)Sprague * Ajwain	Leaves, Root, Fruit	(+)	(-)	(-)	(-)	(+)

Note: (-) = Absent
(+) = Reaction recorded if the addition of the reagent produced only a slight opaqueness
(++) = Reaction recorded if a definite terbidity was observed but no flocculation
(+++) = Reaction recorded if the reagent produced a definite heavey precipitate or heavy flocculation

Acknowledgements

Grateful thanks are due to Dr. K.K. Nag, Professor and Head, Professor N. Dayal, University Department of Botany, Ranchi University, Ranchi. Professor P.C. Tripathy, Utkal University, Bhubneshwar and Dr. V.K. Singh, Reader, P.G. Department of Botany, Co-operative College, Jamshedpur for continuous encouragement and valuable suggestions. Author extends the sincere thanks to University Grants Commission, Calcutta for providing financial assistance.

References

Benjamin, B.D., Siphahimalani, A.T. and Heble, M.R. (1990). Growth and alkaloid production in different cell lines of *Catharanthus roseus* (L) G.Don. Ind. J. Exp. Biol. 28(6): 516-518.

Bhattacharjee, A.K., and Das, A.K. (1969). Quart J. Crude Drug Res. 9: 1408.

Dhar, M.M. and Bhattacharji, S. (1955). A method for the estimation of reserpine in samples of *Rauwolfia* alkaloids. J. Sci. Industr. Res. 14(B): 276-278.

Dutta, N.K. and Iyer, S.N. (1968). Antiamoebic value of berbernine and Kurchi alkaloids. J. Ind. Med. Assoc. 50: 349.

Eilert, U., Deluca, V., Kurz, W.G.W. and Constable, F. (1987). Alkaloids formation by habituated and tumorous cell suspension cultures of *Catharanthus roseus*. Plant Cell Rep. 6(4): 271-274.

Jai Shankar and Basu, N.K. (1961). Separation of Kurchi alkaloids by paper chromatography. Proc. Rajasthan Acad. Sci. 8: 88.

Lin-Geeng Chen, Ling-Ling Yang, Kum Ying Yen, Tsuhomu Hatano, Takashi Yoshida and Takuo Okuda. (1995). Tannins of Euphorbiaceae plant XIII New Hydrolyzable Tannins Haring phloroglucinol Residue from *Glochidion rubrum* Blume. Chem. Pharm. Bull. 43(2): 2088-2090.

Mahato, A.K., Ram, R.L. and Mahato, Pushpa. (1992). Ethnobotanical Wealth of Chhotanagpur Plateau-I: Some medicinal plants used in dysentery. Geobios New Reports. 11(2): 123-124.

Mukhopadhyay, S., Sayed, A.E., Handy, G.A., Cordell, G.A. (1988). *Catharanthus* alkaloid XXXVII, lb-epi-z-iso sitsirikine a monomeric indole alkaloid with antineoplastic activity from *Catharanthus roseus* and *Rhazya Stricta*. J. Nat. Prod. 46: 409-413.

Prasad, S. and Kaul, P.N. (1957). Seasonal variation in the alkaloidal content of Kurchi bark. Indian J. Pharm. 19: 131.

Ram, R.L. and Saha, Vineeta. (1998). Ehtnobotanical Wealth of Ranchi district, Bihar Part-I: Herbal medicinal plants used against dysentery. Ad. plant Sci. 11(1): 249-252.

Ram, R.L. and Saha, Vineeta (1998). Ehtnobotanical Wealth of Ranchi district, Bihar Part-II: Herbal medicinal plants used against dysentery. Bionature. 18(1): 13-15.

Ruyter, Carl. Michael, Mahmood, Akram, Ihan Illahi and Joachin Stoeckigt. (1991). Investigation of the alkaloid contents of *Rauwolfia Serpentina* roots. Planta Med. 57(4): 328-330.

Smith, J.L., Smart, N.J., Kurz, W.G.W. and Misawa. (1987). The use of organic and inorganic compounds to increase the accumulation of indole alkaloids in *Catharanthus roseus* (L) G. Don. Cell suspension culture. J. Exp. Bot. 38(194): 1501-1506.

Takashi Yoshida, Koh-ichi Yokoyama, Osamu Namba and Takuo, Okuda (1991). Tannins and related Polyphenols of Euphorbiaceous plant VII. Tirnecullins A, B and Euphorbin F., Monomeric and Dimeric Ellagitamins from *Euphorbia trirucalli* L. Chem. Pharm. Bull. 39(5): 1137-1143.

Takashi Yoshida, Osamu Namba, Kenji Kurokawa, Yoshiaki Amakura, Yan-ze. liu and Takuo, Okuda. (1994). Tannins and related polyphenols of Euphorbiaceous plants. XII. Euphorbins G. and H., New Dimeric Hydrolyzable Tannins from *Euphorbia prostrata* and *Euphorbia makinoi*, Chem. Pharm. Bull. 42(10): 2005-2010.

Yoshiaki, Amakura and Takashi Yoshida. (1996). Tannins and related polyphenols of Euphorbiaceous plants. XIV. Euphorbin I.a new Dimeric hydrolyzable Tannin from *Euphorbia watanabei* Chem. Pharm. Bull. 44(7): 1293-1297.

18

Drinking Water Quality of Koshi Zone (Bihar) and its Impact on Population Health

B.N. Pandey, P.K.L. Das, A.K. Jha and A.K. Ojha
P.G. Deptt. of Zoology,
Purnia College, Purnia - 854 301, Bihar

Abstract

A preliminary investigation on the physico-chemical quality of drinking water of Koshi zone was studied. The results indicate that the vast majority of water resoruces possess acceptable physical and chemical qualities (except iron and nitrate). The bacteriological analysis revealed that water in the study area is not suitable for human consumption. All waters presented evidence of faecal contamination from human and/or animal origin. Due to use of such contaminated water the inhabitants of the area suffer from a variety of water borne dieseases like diarrhoea, cholera, dysentery and typhoid fever etc. On the whole, the water is not suitable for drinking purpose. It was observed that the main causes of pollution of ground water are high level of water, dumping of sewage near the water source, location of hand pumps/wells nears agricultural fields/cattle sheds and lack of sanitation etc.

Introduction

Ground water is about 95 per cent of the total available water all over the world. Ground water is used for various purpose like drinking, livestock, agriculture etc. It is, however, gradually getting polluted due to various human activities. Ground water pollution may be defined as the artificially induced degradation of natural quality of water. The qaulity of drinking water highly depends upon its sources and varies according to geo-physical quality of soil. Due to direct or indirect interferences of human beings, the waters have been found to contain various pollutants that cause harmful effect on drinking waters and adverse impact on human health and aquatic life as well (Telliard and Rubin, 1987). Slight variation in physico-chemical properties of drinking water influence the population health up to a great extent as well as socio-economical condition. It is estimated that every year two million deaths occur in India due to water borne diseases and more than 50 million persons are partially incapacitated annualy due to these ailments. Sapru (1989) pointed out that safe drinking water constitute only two per cent of the earth's total water supply. Ramana (1980) indicates that 95 per cent of the world's water resources are saline and only 5 per cent are freshwater. Of the freshwater 98 per cent is ground water and 2 per cent surface water. It is further observed that access of safe drinking water is quite inadequate and also unevenly distributed. In India, only 12 per cent people get good clean drinking water (Kudesia, 1986).

Different kinds of diseases are transmitted through contaminated water and poor sanitation. The diseases are caused by infective agent such as Bacteria, Protozoa, Viruses or Worms. The implication due to lack of safe drinking water for child health is clear. An estimated 8.7 per cent of the deaths in the old age group and 17.1 per cent of the deaths between 5 to 15 years age group are due to water borne diseases alone *i.e.* diarrhoea. An estimated 5 million infants die in different parts of the world every year.

The utilization of ground water for domestic, industrial and irrigation purposes has been contantly increasing. Though ground water is a replenishable resource of water which has inherent advantages over the surface waters. It has some limitations on its utilizable potential. Rapid urbanization, industrialization and mining activities have resulted in changed physico-chemical properties of the ground water. The general scarcity of safe potable water has become a global problem and the U.N. Water Conference held at Mar Plata (Argentina) in 1979 designated the period from 1981 to 1990 as the International Drinking Water Supply and Sanitation Decade. Inspite of launching various programmes during this period by the Government of India and the State Government, the major population of Bihar still utilize untreated ground water for drinking and cooking purposes of 22 towns (56.51 per cent) in

North Bihar and 49 towns (78.27 per cent) in South Bihar depend solely upon the underground water resources (Ghose *et al.*, 1986). Much of the ill health in the underdeveloped countries is largely due to the lack of safe drinking water. Water pollution is by for the most serious in its implications for the health of the nation. WHO is of the view that, one in every three people in the world lacking clean water is an Indian. According to one estimate two-third of all illness in India are related to water borne diseases such as typhoid, infective hepatitis, cholera, diarrhoea and dysentery. Almost 73 million workdays are lost due to water borne diseases. Estimates show that water pollution causes around two million deaths in India every year (Sapru and Bhardwaj, 1990).

Studies aimed to determine the quality of underground water in different parts of the India have shown that ground water contains many chemical species, whose type and concentration depend upon the surface and subsurface environments. Various studies (Ali and Iqubaluddin, 1992; Chacharkar *et al.*, 1992; Dayal and Singh, 1991; Olaniya and Saxena, 1977; Jais *et al.*, 1993; Somani *et al.*, 1972; Gupta and Nathawat, 1991; Kishore *et al.*, 1990; Kudesia *et al.*, 1996) carried out in last years have clearly shown that underground water is also becoming contaminated with hazardous substances and pathogenic micro-organisms. Ingestion of this polluted water can result in serious health problems (Mukherjee *et al.*, 1995 and Shrivastava and Choudhary, 1997).

The Study Area

North Bihar is inhabited by scheduled castes and migrants scheduled tribes along with backwards and other well to do castes. In North Bihar (Koshi zone) the main sources of drinking water are the hand pumps and wells. However in some areas river water is also used for drinking purpose. Hand pumps and wells are swallow type as water level is very high in the area. Shallow hand pumps and wells are notoriously liable to be polluted by neighbouring sources of contamination like latrines, urinals, drains, soak pits and dumping waste products etc. Hand pumps as well as wells therefore put forward health hazards to the community. The preliminary survey suggests that the poor water quality is one of the main reasons of the ill health of the populations. Some sources also contain Bacteria, Phytoplanktons and Zooplanktons. High iron contents is the characteristics feature of drinking water of this area.

Water borne diseases are very common in the area, which was once popularly known as the "Kala Pani" of the Bihar and still it is a backward area and no systematic study has been carried out dealing with the quality of the drinking water in relation to its impacts on health of the populations as well as socio-economical variables.

Material and Methods

Water samples from different areas were collected and analysed for different parameters according of APHA (1975) and Trivedy and Goel (1984). Ten water samples were collected from each study area.

Experimental work was conducted for the bacterial and its potability. Sterilized tubes were used for collection of water samples for bacterial count.

Results and Discussion

The physico-chemical characteristics of ground water in different villages show significant variations in the ground water chemistry depending on the geology, seasonal and climatic conditions and physical and biological activities in the region.

The pH of water samples varies between 6.84 - 7.42, which is well within the safe limits prescribed by different agencies (WHO and ISI). It is known that pH of water does not cause any severe health hazard. Turbidity as insoluble particulates imparts to passage of light through water by scattering and absorbing the rays. Turbidity measurements is important from the aesthetic point of view. A water is turbid when it contain visible materials in suspension. Turbidity in water has a significant effect on the micro-biological quality of drinking water due to the presence of bacteria and virus. The higher value of trubidity and total solids of potable water may create taste, odour and colour problems. It may also cause throat and gastrointestinal problem (ISI, 1983 and WHO, 1984). In the present investigation turbidity is within the permissible limits. The total hardness is an important parameter of quality of water, whether it is to be used for domestic or industrial purpose. Depending on the interaction with other factors such as pH and alkalinity, water with hardness above approximately 200 mg/l may cause scale deposition in the distribution system and results in excessive soap consumption and subsequent scum formation. Soft water with hardness less than 100 mg/l may have low buffer capacity and more corrosive for water pipes (WHO, 1984). Hardness has no known adverse effect on health. However, its maximum permissible level has been prescribed for drinking water as 500 mg/l by WHO. According to same classification water having hardness up to 75 mg/l is classified as soft. 76-150 mg/l as moderately soft, 151 to 300 mg/l as hard and more than 300 mg/l very hard (Twort *et al.*, 1974) on the basis of this the water of the present investigation may be considered as moderately soft. Hardness of the water of study area ranges from 128 mg/l to 164 mg/l which is within permissible limits. TDS is an important parameter for drinking water and water to be used for other purposes. Beyond certain limit it imparts a peculiar taste to water and reduce its portability. TDS in water less than 1000 mg/l is classified as non-saline. In the present investigation the value of total dissolved solid is within the permissible limits. Presence of sulphate has less effect on the taste of water compared to the presence of chloride. The sulphate contents of natrual water is an important consideration in determining their suitability for public water supplies. Excess Na_2SO_4 and $MgSO_4$ can cause cathartic action. Higher concentration of Na_2SO_4 in water can cause misfunctioning of alimentary canal. In the present investigation the sulphate value varies between 30-68 mg/l which is within the permissible limits.

Fluoride is an essential element which is required in trace amount for all human beings and is one of the normal constituents of all diets. It commonly occurs in the earth's crust as fluospar, cryolite and fluorapatite and ranks thirteen among the elements in border of abundance (Bhargava *et al.*, 1978). If fluoride in drinking water is less than 0.5 mg/l, the incidence of dental caries is likely to be high. When present in much greater concentration it can cause endemic cumulative fluorosis with resulting skeletal damage. Further, with increase of fluoride content in drinking water, there is mottling of teeth, osteosclereosis, thyroid disorder and kidney damage and very high concentration of fluoride when ingested may prove fatal. The prescribed limit for fluoride in drinking water is 0.5 to 1.2 mg/l and the maximum permissible limit of fluoride is 1.5 mg/l. Fluoride has considerable physiological importance for man and animals. If one is exposed for longer durations to fluoride it may pose serious health hazards. The major health problems caused by fluoride are dental caries, teeth mottling, skeletal fluorosis and deformation of bones in children as well as in adults.

Recent researches by Geological Survey of India have laid to the discovery of a natural mineral set pen-time which can absorb large quantities of fluorides from waters and can act as a defluorising agent with considerable success in the treatment of fluorosis (Panda and Singh, 1997). In the present investigation the value of fluoride ranged from 0.6 - 0.9 mg/l which is however, below the permissible limits.

Chloride in the form of chlrodie ions, is one of the major inorganic anions in water. Chloride in natural water may sometimes be due to leaching of rocks. Human body releases a very high quantity (6gm/person/day) of chloride. High concentration of chloride gives an undesirable taste to water and beverages. Lockhard *et al.*, (1955) have reported that the thresholds for chloride content of water may cause unpleasant taste of coffee. In potable water, the salty taste produced by chloride concentration is variable and depends on the chemical composition of water. The chlordie value in the present investigation ranged from 2Q-170 mg/l which is much below the permissible limits.

Arsenic is an ubiquitous element present in various compound throughout the earth's crust. In 1556, Agricola reported that some health effects might be related to arsenic exposure. A few epidemiological studies revealed that skin, lung, liver and bladder cancers were associated with arsenic exposure *via* inhalation or ingestion (Bates *et al.*, 1992; Enterline *et al.*, 1995). Inhaled or ingested arsenic causes cancer in humans but not in other species (Bates *et al.*, 1995). However, in the present investigation arsenic was found nil in all the studied sites.

As the concentration of TDS, sodium and chloride in ground water has a bearing on the human health, a detailed discussion is needed in this respect. Drinking water standards are generally based on two main criteria (Davis and Dewiest, 1966) : (1) presence of objectionable taste, odour and colour and (2) presence of substances with adverse physiological (health effects) characteristics. As far as first criterion is concerned, none of the water samples has objectionable colour and odour. For the second criterion to be examined, the chemical analysis results were taken into consideration.

Contamination of ground water with nitrate is the cause of one of the most serious impact of human activities on hydro-geological systems (OECD, 1986; Jenny and Daniel, 1995). Since high concentration of nitrate in drinkig water may have adverse health effects (Frazer *et al.*, 1980), ICMR (1975) and WHO (1984) have proposed guideline value of 45 mg/l for drinking water, concentration that is approached or exceeded under diverse agricultural management systems. Since nitrate is mobile in most of the soils, leaching losses of nitrate from fertilized soils and also cause economic implications due to high cost of agricultural fertilizers.

There have been numerous studies relating to nitrate contamination of ground water in India and abroad (Lakshamanan and Krishna Rao, 1986; Pandey *et al.*, 1986; Andamuthu and Subburam, 1994). In India nitrate (as NO_3) as high as 530 mg/l has been reported in Churu district of Rajasthan. Andamuthu and Subhuram (1994) have reported that 36.43 per cent of wells in Periyar district have nitrate more than the maximum limits prescribed by WHO (1993). Gupta and Saxena (1997) found higher concentration of nitrate in ground water of Agra. Ground water pollution due to sewage percolation in Jaipur have been reported by Olaniya and Saxena (1977). Increasing rate of nitrate in ground waters is an alarming problem these days because it is associated with a disease known as methemoglobinaemia (infant cyanosis), if nitrate is present in concentration above 45 mg/l (WHO, 1984). Researchers have linked nitrate in drinking water with recurrent acute respiratory disorders in children. Also there are certain possible links between high nitrate consumption and certain type of cancers (I.S.C., 1974 and Davies, 1980) because in stomach nitrates may form N-nitroso compounds as carcinogens in stomach. The nitrate values in water samples of Taranagar, Ramnagar and Barahkona are higher.

The high value of nitrate in some study areas is due to (a) irrational location of wells and hand pumps with respect to agricultural fields where men and women go for morning evening defecation, (b) low level of water and unprotected wells, (c) another significant cause of nitrate pollution in the area may be attributed to sewage disposal, as planned sewage disposal systems are lacking in the area. The main sources of drinking water in these villages are hand pumps and wells. It is to be noted these villages are located nearby to agricultural fields with no drainage facilities. Most of the domestic sewage is found accumulated near the hand pumps. Fairchild (1987) has reported that nitrate contamination of ground water is due to increasing use of nitrogenous fertilizers. NAEP (1991) and J.C. German Management System (1989) have concluded that residual nitrate in the soil is the major cause of nitrate contamination in ground water.

Abnormal high concentrations of nitrates in some of the study area are mainly associated with land use and human activities. The value of nitrate exceeds the permissible limits for drinking purposes, causing serious health problem to the local inhabitants. Several measures should be implemented to ensure the quality of resources—

(i) Regular water quality monitoring

(ii) Balanced use of fertilizers with proper drainage outlets

(iii) Modernisation of sewage disposal system.

Heavy metals are those, whose density is 5 or more. When these heavy metals are in higher concentration in the ground waters they cause pollution and can get access through food chain to humans and animal kingdom. When such contaminated water is consumed over a period of time, it can cause serious health hazards. Heavy metals are usually present in trace amounts in natural waters but many of them *viz.* Cd, Cu, Cr, Mn, Pb and Fe are quite toxic to human health even at very low concentration (Verma *et al.*, 1995). Some of the heavy metals in excess may cause serious physiological disorders.

Sodium concentration is another important factor in classifying irrigation water because, sodium reacts with soil to reduce its permeability. Excessive sodium in irrigation water may create toxicity to certain sensitive crops. Willicox (1955) relates the Na, K and Mg concentration. Sodium and potassium are naturally occurring elements in ground water. Industrial and domestic wastes also add sodium to the ground water. Sodium concentration more than 50 ppm makes the water unsuitable for domestic use. Besides the presence of arsenic and lead are also known toxic for the human body. High intake of copper may results in damage to liver. Still there are many heavy metals which are toxic to human. Lead is known to inhibit the function of certain enzymes necessary for the formation of the Heam in bone marrow, the pigment that combines with protein to form haemoglobin (Pillai, 1983). In the present investigation lead, arsenic and copper were found to be absent. However, level of potassium, sodium, zinc, clacium, magnesium, cobalt and cadmium were found to be within the permissible limits.

Iron is a widely encountered contaminant in both ground and surface waters. However, iron remains usually higher in ground waters as compared to surface waters. It has been reported that presence of excess quantities of iron will produce distressing gastrointestinal disturbances (Babcock, 1951 and Sharmah, 1994). Excess amount of iron, which is absorbed is stored in liver, spleen and bone marrow, causes weakness, aging and other health problems. Iron has got a microbiol significance since a few micro-organisms, such as *Crenothrix, Leptothrix* and *Gallionella* are able to utilise dissolved iron as an energy source and convert ferrous into ferric hydroxide. Higher iron value may be related to agricultural activity in the catchment areas and low water levels. Pandey *et al.*, (1992 and 1994) have reported the high percentage of iron and nitrate in some areas of Koshi zone.

Deficiency of iodine is the common feature of the Kosi zone. Iodine deficiency disorders are the leading cause of preventable brain damage in the foetus and infants as well as of retarded psychomotor development in young children. Its deficiency is also responsible for stillbirths, abortions, congenital anomalies, endemic cretinism, mental disorders and endemic goitre (Hetzel, 1993). Of these goiture, stillbirth and cretinism have been found common in the Purnia district of Bihar (Pandey *et al.*, 1999). In the present investigation iodine content varies from 0.001 to 0.050 mg/l which is quite less than permissible limits.

Since water borne infections are due to faecal pollution of the water, the detection of such contamination in water becomes the prime object of the bacteriological examination of water. The most important parameter of drinking water quality is the bacteriological quality *i.e.* the content of bacteria and viruses. The bacteriological examination of water has a special significance in pollution studies as, it is a direct measurement of deleterious effects of pollution on human health. The great danger to health is due to the presence of excremental bacteria as contaminated water may convey the causative organism of diseases.

According to the United States National Research Council, one objective of the physical and chemical treatement of water intended for public consumption, is to reduce the level of the total coliform less than 1 coliform per 100 ml. Such treatment has proven to be an effective barrier against transmission of water borne diseases (NAS, 1982). The presence of heterotrophic bacteria can be used to predict the presence of total coliform. Heterotrophic bacteria count and coliform enumeration of water (Edberg and Smith, 1989). *E. coli* is the most numerous coliform in human and animal intestine and is derived almost exclusively from these sources. The presumptive coliform count laid out in the area of study indicates that the water contaminated with faecal matter and thus not potable. The results indicate that none of the areas were without contamination. Mueller (1979) found *Salmonella sp.* (4/100 ml) and Escherichia coli (408/100 ml) in drinking water. Sabwa and Githeko (1985) have reported that *E. coli* ranged from 20 to 1800/100 ml in potable water. Coliform MPN/100 ml of a sample in most of the areas is very high which indicates a poor quality of water and possible fear of sewage getting mixed with the water source. The depth of the tube wells ranged from 15-20'. However, it has been reported that the bacteriological quality of water did not depend on the depth of the tube wells. Even deep tube wells have shown faecal pollution (Joshi *et al.*, 2000). In most of the study areas, the sewage is carried by Nalas which are not cemented. Thus, overall faulty drainage system appears to be the major source of faecal contamination. The study reveals that sewage contamination has badly affected the hydro-eco system and posses a serious health hazard.

Sources of Ground Water Pollution

(i) Higher level of water (10'-20').

(ii) Dumping of sewage near the water source which contains organic matters and pathogenic agents.

(iii) Location of hand pumps near agricultural fields due to which fertilizers and pesticides used in agriculture contaminate the water source through seepage.

(iv) Lack of proper sanitation.

(v) Bathing and washing of clothes and utensils near the hand pumps.

(vi) Location of hand pumps/wells near the cattle sheds.

(vii) The habit of to make water and morning and evening defecation by men and women near the hand pumps/ wells.

(viii) Water disposal in excavations.

(ix) Waste disposal in the well excavations.

Water Borne Diseases

Water contamination is an integral factor, which has cognizably increased with urbanization and industrialisation by direct and indirect means. This posses a threat to water sources, which have evolved into a formidable factor in the spread of human and animal diseases. The causative agents for typhoid fever, amoebic dysentery, cholera and para-typhoid fever are spread by water, (Charles and Alice, 1957). It is estimated that every year two million deaths occur in India due to water borne diseases and more than fifty millions persons are partially incapacitated annually due to these ailments.

Due to poor quality of drinking water, the population of the surveyed area suffer from a variety of water borne diseases has been shown in Table 18.2.

All water showed faecal contamination and should be treated before using for drinking purpose. The high incidence of iron and nitrate in some samples may cause serious health effects. Since, in this area, the main source of drinking water is the hand pump, there is an urgent need to design a health education programme with respect to environmental sanitation and safe drinking water supply. Hand pumps and wells should be located at higher elevations and at safe distance from agricultural fields. There should be no sewage disposal near hand pumps and wells. Sewage disposal facilities should be at least 50′-60′ away from the source of drinking water. Microbiological study of water should be carried out at regular intervals. The establishment of minimum standards of physical, chemical and bacteriological quality for public water supply is of fundamental importance in achieving the objective of insuring safe drinking water to the community.

To control diarrhoeal diseases, it is essential that drinking water is protected from poullution with faecal bacteria, this is only possible if people use latrines and wash their hands with soap after defecation and before eating.

Table 18.1: Analytical Investigation on Potable Water quality of Purnia District

Study Sites	*Colour*	*Turbidity*	*pH*	*Total Hardness*	*Alkalinity*	*Fluoride*	*Chloride*	*Nitrate*	*Sulphate*	*TDS*	*Iron*	*MPN*
				mg/l	*mg/l*	*mg/l*	*mg/l*	*mg/l*	*mg/l*	*mg/l*	*mg/l*	*mg/l*
Khunkhuni	Clear	6	7.36	140	156	0.6	36	4	40	200	4.0	95
Geria	Clear	10	7.30	142	170	0.8	150	50	65	212	5.9	98
Ramnagar	Clear	10	6.84	138	144	0.6	40	42	60	210	4.5	80
Taranagar	Clear	5	7.04	152	160	0.7	170	85	50	215	5.6	60
Rakahkona	Clear	6	7.32	128	132	0.7	20	20	30	218	6.0	98
Maranga	Clear	9	7.42	145	175	0.9	160	71	68	225	6.8	84
Dharhara	Clear	10	7.35	135	145	0.9	170	52	45	218	1.5	52
Court Stn.	Clear	4	7.26	160	162	0.8	128	30	40	240	0.2	70
Makhnaha	Clear	6	7.12	164	155	0.6	150	61	48	222	1.8	64
Ranipatra	Clear	7	7.16	136	158	0.6	68	48	55	275	6.7	62
Gandhi Nagar	Clear	8	7.12	130	153	0.7	60	32	48	206	4.9	58
Champa Nagar	Clear	6	7.25	148	152	0.6	70	3.8	46	208	5.4	64

Table 18.1 contd....

Study Sites	*Arsenic*	*Copper*	*Lead*	*Iodine*	*Magnesium*	*Zinc*	*Potassium*	*Sodium*	*Calcium*	*Cadmium*
Khunkhuni	0	0	0	0.002	17.55	0.182	18.21	29.89	33.66	0
Geria	0	0	0	0.002	7.70	0.0195	18.28	30.40	34.40	0
Ramnagar	0-	0	0	0.001	14.42	0.141	16.60	31.80	32.60	0
Taranagar	0	0	0	0.000	13.52	0.142	16.80	33.20	34.10	0
Darahkona	0	0	0	0.005	17.54	0.171	15.60	34.40	33.80	0
Maranga	0	0	0	0.020	15.21	0.172	16.70	30.80	33.70	0
Dharhara	0	0	0	0.010	12.40	0.162	18.20	34.80	32.40	0
Court Stn	0	0	0	0.004	14.10	0.157	17.50	35.40	31.80	0
Makhnaha	0	0	0	0.002	13.21	0.142	16.24	36.21	32.50	0
Ranipatra	0	0	0	0.001	14.41	0.151	19.40	40.50	33.20	0
Gandhi Nagar	0	0	0	0.050	15.29	0.148	17.60	38.90	36.20	0
Champa Nagar	0	0	0		14.01	0.149	18.70	37.24	38.10	0

Table 18.2: Prevalence of Water Borne Diseases among populations of Koshi zone

Population	*Number*	*Diarrhoea*	*Jaundice*	*Hepatitis*	*Amoebic Dysentery*	*Typhoid*
Brahmin	266	152 (19.54%)	41 (15.41%)	10 (3.75%)	102 (38.34%)	79 (29.69%)
Kulhaiya	265	180 (7.92%)	101 (38.11%)	21 (7.92%)	201 (75.84%)	98 (36.98%)
Mushar	260	170 (65.38%)	46 (17.69%)	17 (6.53%)	139 (53.46%)	88 (33.84%)
Kharwar	190	76 (40.00%)	36 (18.94%)	18 (9.47%)	121 (64.21%)	94 (49.47%)
Kisan	198	60 (30.30%)	43 (21.71%)	15 (7.57%)	117 (59.03%)	88 (44.44%)
Lohra	129	35 (27.13%)	30 (23.25%)	11 (8.52%)	89 (68.99%)	71 (55.03%)
Munda	245	95 (38.77%)	49 (19.99%)	15 (6.12%)	130 (53.06%)	73 (29.79%)
Oraon	255	57 (22.35%)	38 (14.90%)	10 (3.92%)	103 (40.39%)	80 (31.37%)
Santal	282	65 (23.04%)	52 (18.43%)	12 (4.25%)	104 (36.87%)	78 (27.65%)
Chamar	192	45 (23.43%)	38 (19.79%)	8 (4.16%)	79 (41.14%)	36 (18.75%)

References

Ali, A. and Iquabaluddin (1992). Ground water quality in parts of Gomati Basin, District Sultanpur (U.P.). J. Indian Water Resources Soc., 12(182): 54-60.

Andamuthu, R. and Subhuram, V. (1994). Nitrate concentration in ground water of LBP main canal command area in T.N. IJEP, 14(8): 462-467.

APHA (1975). Standard methods of examination of water and waste water 14th ed. APHA, Washington, USA.

Babcock, R.H. (1951). Iron and Manganese in water supplies and methods of removal. Water Sewerage Work. 98(10): 442.

Bates, M.N., Smith, A.H., Hopenhayn-Rich, C. (1992). Arsenic ingestion and internal cancers : A review. Am. J. Epidemiol. 135: 462-476.

Bates, M.N., Smith, A.H., Cantor, K.P. (1995). Case control study of bladder cancer and arsenic in drinking water. Am. J. Epidemiol. 141: 523-529.

Bhargava, R.K., Saxena, S.C. and Thergaonkar, V.P. (1978). Ground water quality in Ajmer district. Ind. J. Environ. Hlth. 290-299.

Chacharkar, M.P., Madhawan, A.K., Singh, D., Soni, N.K. and Bairwa, S.P. (1992). Some rare elements in the ground water of Rajasthan. Indian J. Environ. Prot., 12(6): 445-450.

Charles, F.C. and Alice, C.S. (1957). Principles of Microbiology. The C.V. Mosby Company, St. Louis. pp. 579-583.

Davies, J.M. (1980). Stomach cancer mortality in workshop and other Nottinghamishire mining towers. J. Cancer, 41: 438-445.

Davis, S.N. and Dewiest, R.J.M. (1966). Hydrogeology. John Wiley and Sons, New York pp. 463.

Dayal, G. and Singh, R.P. (1991). Heavy metal contamination of ground water in Agra city (U.P.). Proc. Nat. Acad. Sci. India. 61(A): 569-572.

Edberg, S.C. and Smith, D.B. (1989). Absence of association between total heterotrophic and total coliform bacteria from a public water supply, Appl. Environ. Microbiol., 55(2): 380-384.

Enterline, P.E., Day, R. and Marsh, G.M. (1995). Cancers related to exposure to arsenic at copper smelters. Occup. Environ. Med., 52: 28-32.

Fairchild, D. (1987). Ground water quality and agricultural practices, Lewis Publ., Chise, Mi. pp. 151-174.

Frazer, P., Chilvers, C., Beral, V. and Hill, M.J. (1980). Nitrate and human cancer: A review of the evidence. Int. J. Epidemiol., 9: 3-11.

Ghose, N.C., Sharma, C.B. and Sinha, S.N. (1986). A appraisal of the quality of drinking water resources in Bihar. A case study from Eastern India. J. Landscape. Systems abd. Ecol. Studies, 9(1): 55-63.

Gupta, A. and Nathawat, G.S. (1991). Heavy metal pollution of the surface and drinking water of Sanganer. Acta. Ecological, 13(1): 11-15.

Gupta, A.K. and Saxena, G.C. (1997). Nitrate contamination in ground waters of Agra and its correlation with various water quality parameters including heavy metals. Poll. Res., 16(3): 155-157.

Hetzel, B.S. (1983). Iodine Deficiency Disorders (IDD) and their eradication. Lancet., 1: 1126-1129.

ICMR (1975). Manual of standards of quality of Drinking Water Supplies. Special Report Series No.44 IInd Ed.

I.S.C. (1974). A report on water quality and treatment of nitrates in water supply. Aqua, 1: 5-24.

ISI Report (1983). Standard specification for drinking water. Indian Standard Institution, New Delhi publication no.10500.

Jais, G.K., Shrivastava, R.M. and Verma, N.K. (1993). Chemical quality of surface and ground water around Vijaypur (District Guna) M.P., India, Ultra Scientist Phyl. Sci., 5(2): 207-209.

J.C. German Management System (1989). J.G. Ger. Management Sys. to reduce nitrates. Eisevier Applied Science.

Jenny, S. Reynolds and Daniel D. Richter. (1995). Nitrate in ground water of the Central Valley Costarica. J. Envi. Inter., 21(1): 71-79.

Joshi, S., Verma, S., Chitnis, V., Hemvani, N., Trivedi, R., Ravikant and Chitnis, D.S. (2000). Bacterial contamination of underground water in Indore city. J. Env. and Poll., 7(4): 309-312.

Kishore, K., Singh, M., Singh, N.K. and Dutta, G.R. (1990). Evaluation of some aspects of abiotic factors for drinking water resources of Jamshedpur township. J. Freshwater. Biol. 2(3): 285-288.

Kudesia, V.P. (1986). Water pollution, Pragați Prakashan, Meerut.

Kudesia, V.P., Verma, S.P., Singh, K.P. and Sanjeev (1996). Pollution studies of drinking water qaulity of environmentally degraded village Kamalpur (District Meerut) U.P. and their remedial measures. Indian. J. Environ. Agricl. 1(1): 38-44.

Lakshamanan, A.R. and Krishna Rao, T. (1986). Nitrate and Fluroide levels in drinking water in town city of Hyderabad and Secunderabad, I.J. Env. Hlth., 28(1): 45.

Lockhard, E.E., Ticker, C.L. and Merritt, M.C. (1955). The effects of water impurities on the flavour of brewed coffee food Res., 20: 698-705.

Mukerhjee, S., Pal, O.P. and Pandey, A.K. (1995). Case studies on sporadic fluoride contamination in ground water, district Unnao. Bhu-Jal News (Quarterly). J. of Central Ground Water Board, 10(1): 1-6.

Mueller, H.E. (1979). The occurrence of Salmonella in drinking water. Zentralbl. Bakteriol. Parasitenkd. Infection. Skr. Hyg. Erste. Abt. Orig. Reihe. Bhyg. Mrankena-Haushyg. Betri Edshyg. Prac. Med., 169(5/6): 551-559.

NAEP (1991). Environmental impacts on nutrient emissions in Denmark. Publ. NAEP, Denmark, No.1.

NAS (1982). Drinking Water and Health. Safe Drinking Water Committee. National Academy of Sciences, Washington, D.C.

OECD (Organization for Economic Cooperation and Development) (1986). Water pollution by fertilizers and pesticides. OECD, Paris, France.

Olaniya, M.S. and Sexena, K.L. (1977). Ground water pollution by open refuse dumps of Jaipur. Ind. J. Env. Health. 19: 170-188.

Panda, N. and Singh, B.C. (1997). Study of major inorganic anions in various sources of drinking water in port city of Paradeep. Poll. Res., 16(4): 251-253.

Pandey, B.N., Das, P.K.L. and Jha, A.K. (1992). Physico-chemical analysis of drinking water of Purnia district, Bihar. Acta. Ecol., 14(2): 108-114.

Pandey, B.N., Das, P.K.L., Jha, A.K., Tripathi, R.N. and Pandey, S.K. (1994). An assessment of quality of drinking water of Katihar, North Bihar, Acta. Ecol. 16(2): 144-149.

Pandey, B.N., Mishra, S.K., Yadav, S. and Sharma, P.D. (1999). Nutritional deficiency disorders among population of Purnia district, Bihar, Acta. Ecol., 21(2): 55-67.

Pandey, S.P., Hasan, M.J. and Saxena, K.L. (1986). Nitrate, Nitrites in the environment : An overview. I. J. Env. Health, 18(3): 201-206.

Pillai, K.C. (1983). Heavy metals in aquatic environment in : Water Pollution and Management (C.K. Varshney ed), pp. 74-93. Wiley Eastern Ltd., New Delhi.

Ramana, D.V. (1980). An overview of development in Asia Pacific. An UNAPDT Publication. p. 137.

Sabwa, D.M. and Githeko,.A.K. (1985). Faecal contamination of urban community water supplies and its public health implication. Est. Abr. Med. J., 62(11): 794-801.

Sapru, R.K. (1989). Water Pollution Management in Chandigarh. Ind. Jr. of Publ. Adminst. No. 3: 572-581.

Sapru, R.K. and Bharadwaj. (1990). The New Enviornmental Age.

Sharmah, A. (1994). A study on the presence of iron in rural water sources and its removal. B.Tech. Project Report NERIST, Arunachal Pradesh.

Shrivastava, R. and Choudhary, B. (1997). Drinking water quality in an average Indian city : A case study of Agra (U.P.). Poll. Res., 16(1): 63-65.

Somani, I.L., Gandhi, A.P. and Paliwal, K.V. (1972). Note on the toxicity in well water of Nagpur and Jaipur districts of Western Rajasthan. Indian J. Agric. Sci., 42(8): 752-754.

Telliard, W.A. and Rubin, M.B. (1987). Control of pollutants in waste water. J. Chromat. Sci., 25: 322-327.

Trivedy, R.K. and Goel, P.K. (1984). Chemical and Biological Methods of Water Pollution Studies. Enviornmental Publications, Karad.

Twort, A.C., Hoather, R.C. and Law, F.M. (1974). Water supply. Edward Ornold Pub. Ltd., London.

Verma, M.K., Jain, O.P. and Shrivastava, P.K. (1995). Preliminary studies on heavy metals in ground water of Mandi deep by atomic absorption spectro-scopy. Proc. Acad. Environ. Biol., 4(1): 123-126.

WHO (1984). International standards for drinking water. WHO, Geneva.

WHO (1993). Guidelines for drinking water quality (2nd end., Vol.1). World Health Organization, Geneva.

Willicox, L.V. (1955). Classification and use of irrigational water, US Department of Agriculture, Cire, 969 Washington D.D. p. 19.

19

Diurnal Variations in Physico-Chemical Parameters of Bonal Reservoir During Post-Monsoon Season

K. Vijaykumar, Halli Anjinappa, Ramesha. I., Virupanna and C. Padmavathi
Environmental Biology Research Unit,
Department of Zoology,
Gulbarga University, Gulbarga - 585 106, Karnataka

Abstract

The diurnal variations of some physico-chemical parameters were studied from bonal reservoir during post-monsoon season. The pH values decreased while free carbondioxide values increased during night hours. The nutrient concentrations also exhibited a decreasing trend during night hours, but a sharp increase during the day. The release of free carbondioxide as result of respiratory activity of planktonic organisms could be responsible for increase total alkalinity values during night hours.

Introduction

The present study reports the diurnal variation of physico-chemical parameters during post-monsoon season at Bonal reservoir Shorapur Gulbarga, Karnataka state (Information on diurnal variation of physico-chemical parameters from the limnetic water bodies in India are quite substantial (Hegde *et al.*, 1985; Joshi *et al.*, 1989; Tiwari *et al.*, 1989; Vijaykumar and Paul, 1981; Goel *et al.*, 1992 and others). The diurnal variation in physico-chemical characteristics is related directly or indirectly to the solar heating and biological activity. The magnitude and pattern of these variations vary from habitat to habitat, seasons to seasons and from species to species. Ecological models, nutrient budgets and plankton dynamics are often based on monthly or fortnightly samples without any consideration of the changes taking place at short intervals which may, sometimes, be quite marked.

Material and Methods

Bonal reservoir (77°-45′ latitude 16°-33′) is shallow with maximum depth of 3 meters and an area of 275 hectares, is 20 km away from Shorapur Taluk, Gulbarga district Karnataka.

Surface samples for diurnal variation were collected from upper 25 cm station at the interval of 4 hrs. The temperature, pH, Do, Co_2 were determined at the field, while samples were brought to the laboratory and analyzed within 48 hrs after standard methods (APHA, 1985).

Results and Discussion

The observations recorded for 24 hours during post-monsoon season from Bonal reservoir are shown in Table 19.1.

Table 19.1: Diurnal variation in physico-chemical parameters of Bonal reservoir during post-monsoon season

Hours	*Atmos. Temp. °C*	*Water Temp. °C*	*pH*	*DO mg/l*	*CO_2 mg/l*	*TA mg/l*	*PO_4-P mg/l*	*NH_4-N mg/l*
0800	32.00	27.00	8.3	3.6	Traces	235	0.23	0.20
1200	36.00	28.00	8.5	9.6	Traces	370	0.21	0.19
1600	36.50	28.00	8.5	9.8	1.1	375	0.21	0.17
2000	32.00	26.00	8.0	8.0	2.2	235	0.26	0.11
2400	30.00	25.00	7.8	7.8	6.6	235	0.27	0.11
0400	29.00	24.00	7.7	7.7	9.9	280	0.22	0.10
0800	31.00	26.00	8.1	3.8	4.4	240	0.20	0.18

The water temperature followed a definite pattern of diurnal increase (daytime) and decrease (nighttime). Similar findings have been reported by Joshi *et al.*, (1989), Tiwari *et al.*, (1989) Vijaykumar and Pual (1991) and Goel *et al.*, (1992). However, minimum temperature was recorded at morning hours and ranged between 24.0 to 28.0°C. The atmospheric temperature fluctuated between 29.0 to 36.5°C. This shows that the atmospheric temperature varies greatly than the water temperature in all the water resources studied.

Variations in the hydrogen-ion-concentration appear to be more significant in the reservoir values were observed to fluctuate between 7.7 to 8.5. It is evident from the observations that higher values of pH were noticed when free carbondioxide level was low during the daytime, while pH values were noticed low when free carbondioxide was more in the reservoir. This substantiate the view that constant biological input of carbondioxide into the aquatic ecosystem could be primarily responsible for low level of pH as also observed by many workers *viz.* George (1961), Saha (1981), Pandey (1989), Vijaykumar and Pual (1991).

Dissolved oxygen concentration varied between 3.6 to 9.8 mg/l, maximum being recorded at 1600 hours and minimum at 0800 hours. The dissolved oxygen values in the reservoir fluctuated with an increasing trend during the day and a decreasing trend during the night. The enhanced values are obviously due to accelerated photosynthetic activity gradually towards the later part of the day and the lower values during night were as a result of decline in the photosynthetic activity and comparative increase of oxygen consumption for the respiratory activity of the heterotrophs.

The amount of free carbondioxide in water is of ecological importance as it governs the precipitation of calcium in the form of calcium carbonate. The presence of free carbondioxide recorded during the present study indicates that of free carbondioxide which was not utilized during the photosynthesis by the phytoplankton. It varied from traces to 9.9 mg/l. There was a positive relationship between dissolved oxygen and free carbondioxide, which showed a progressive decrease during day and increase during night. The decrease in daytime was evident due to the active process of photosynthesis. Similarly, rise in the free carbondioxide values during night may be attributed to the respiratory activities of the entire biota. It results in lowering the pH as reported by King (1970), Boyd (1972) and Saha (1981).

Total alkalinity varied between 235 to 285 mg/l in the reservoir. The maximum values were recorded at 1600, 1200 and 0400 hours respectively. The increase in total alkalinity in the reservoir during night hours results from release of substantially free carbondioxide during the respiratory activity as also observed by Vijaykumar and Pual (1991).

Ammonical nitrogen and orthophosphate was studied in the reservoir. The lowest concentration of ammonical nitrogen of 0.10 mg/l was noticed at 0400 hours and 0.20 mg/l at 0800 hours. Orthophosphate ranged between 0.20 to 0.27 mg/l at 2400 and 0400 respectively.

References

APHA (1985). Standard method for the examination of water and wastewater, 16th ed., 1268, New York.

Boyd, C.E. (1972). Source of CO_2 for nuisance blooms of algae-weed. Science 20(5): 492-497.

George, M.J. (1961). Diurnal variation in two shallow ponds in Delhi, India. Hydrobiologia. 18: 265-273.

Goel, P.K., Kulkarni, A.Y., Khatavkar, S.D. and Trivedy, R.K. (1992). Studies on diurnal variation in some physico-chemical characteristics and phytoplankton of fresh water polluted pond. Indian J. Environ. Protection, Vol.12, No.7

Hegade, G.R., Bengale, U.D. and Bharathi, S.G. (1985). Diurnal variatoin of physico-chemical factor and phytoplankton in a pond and lake near Dharwad, Karnataka state, India. Camp. Physiol. Ecol. 10: 245-250.

Joshi, S.N., Tiwari, N.C. and Gour, M.R. (1989). Diurnal cycle of abiotic parameters of three natural drinking water resources of Gopeshwar (Chamoli) Environ. and Ecol. 7: 105-108.

King, D.L. (1970). The role of carbon in eutrophication. J. Water Poll. Cont. Fed. 42: 2035-2051.

Pandey, R.S. and Daradiyar, P.K. (1989). Diurnal changes of certain physico-chemical factor of Shivaganga pond of Deoghar, South Bihar, India. Environ. and Ecol. 7: 483-485.

Saha, S.K. (1981). The diel cycle of abiotic factor in thermal streams. Camp. Physiol and Ecol. 6: 193-196.

Tiwari, N.C., Joshi, S.N., Bhatt, A.B. and Singh, H.R. (1989). Diel cycle of physico-chemical features of the Sagar Gadhera from Garhwal Himalaya. Environ. and Ecol. 7(2): 358-361.

Vijaykumar, K. and Ravindra Pual (1991). Diurnal fluctuation in some physico-chemical parameter of tank and moat in Gulbarga, Karnataka. J. Natcon. 3(2): 109-113.

20

Effect of Physico-Chemical Factors on Seasonal Abundance of Rotifera in Chandrampalli Reservoir, Gulbarga District

K. Vijaykumar, Virupanna, C. Padmavathi, Halli Anjinappa and Ramesha, I.
Environmental Biology Research Unit,
Department of Zoology,
Gulbarga University, Gulbarga - 585 106, Karnataka

Abstract

Seasonal abundance of 15 rotifers species studied in relation to physico-chemical factors, such as temperature, transparency, pH, total alkalinty, dissolved oxygen, free carbondioxide, phosphate and ammonical nitrogen on the basis of observations made from June, 1997 to May, 1998 in the Chandrampalli reservoir. Rotifer showed three maxims dominated by *Branchionus* sp., *Asplanchana* sp. And the peaks were recorded in October 1997, March and May, 1998. The abundance and temperature were found to be important factors in governing in their population density. The dissolved oxygen and free carbondioxide were varying from 2.5 to 8.0 mg/l and traces to 8.8 mg/l respectively.

Introduction

The study of zooplankton communities provides ways of predicting and increasing the productivity of lake (Vijaykumar and Basalingappa, 1992). The important of rotifer especially as fish food in the wild and hatchery systems, is common knowledge among hydrobiologist and fish breeders. Rotifera are also being used in pollution studies under experimental situation and as indicators of tropical condition in the field and as important food link in pelagic ecosystem. These organisms in responses to changed climatic condition like other zooplankton are known for their population fluctuation during different season. In India majority of workers have been reported on zooplankton community, specially, regarding their taxonomy, community structure, apart from their phenological events. Some of the important contribution are those of Arora (1966), Nayer (1968), Jone (1973), Sharma and Michael (1980), Chakrabarthy (1990), Singh (1991), Bais (1992), Baruah *et al.*, (1994), Vijaykumar (1996), Vijaykumar *et al.*, (1999). The purpose of study was to describe the effect of different parameters on the abundance of rotifers.

Material and Methods

The Study Area

The Chandrampalli reservoir (17°-21′ N latitude and 77°-25′ E longitude) is a small irrigation water body having a length of about 3,023 ft., a catchment area of about 170 sq miles and having height of 94 ft. It is located on Saranala (Krishna basin) near Chandrampalli village in Chincholi Taluka of Gulbarga distrcict and located above 467 m.s.l.

The collection were made at monthly intervals from June 1997 to May 1998, and concentrated by filtering of water through plankton net of 55 μm mesh size. They were preserved in 4 per cent formalin. Rotifera and other groups were counted and identified as described by Welch (1948), Tonapi (1960), Nidham and Nidham (1962), Michael (1973) and Patil Gouder (1989).

Result and Discussion

Three major groups, *i.e.* Rotifera, Cladocera and Copepoda (Table 20.2) represented the zooplankton community. Rotifera were represented by 15 species of which *Keratella tropica* was recorded in maximum number (38 individuals per liter) during May 1998 and the minimum number observed in the month of February. *Brachionus angularis* was found in the tank throughout the year execpt in the month of February. *Branchionus rubens* and *Asplanchna pilina, Rotatoria neptunia* was the third taxon in the abundance and occurred collection and other species occurred more or less throughout the period of investigation. Cladocera, *Daphina carinata,*

Moina. macrocopa, Moina. flagellata and *Ceridaphina* and *M. flagellata* were found to occur in this tank of these *D.carinata* and *M. flagellata* were dominant in maximum number, *i.e.* 30 individuals per litre and 28 individuals per litre respectively.

Rotiferans are known to inhabit warm waters ranging from 20-30°C (Nasar, 1975) and the present study supports this view. In the present study *B. angularis, B. ruben, Keratella tropics, Asplanchna pilina* and *Rotatoria neptunia* showed its peak during summer (March, April and May).

From the present investigations it can be concluded that pH values ranging from 7.2 to 9.1 which are suitable. The organisms were more when pH was 8.0 to 8.8.

The dissolved oxygen fluctuated between 2.5 to 5.9 mg/l. Abundance of all species in the month of April, May and June, showed that they can thrive well in low oxygen concentration of 2.5 mg/l.

Free carbondioxide did not seem to be the conditioning factor for the production and growth of rotifers. The free carbondioxide concentration fluctuated within almost traces to 8.8 mg/l during the period of investigations. Five species occrued at free carbondioxide concentration of 1.1 mg/l but registered peak in the month of April, May, June. Interestingly they were also found in fairly good number in the month of August and September at concentrations of 8.8 mg/l.

The lowest population density of rotifers during winter season coincides with a substantial decrease in water temperature. Decline in the population density of rotifers was obvious due to disappearance of *B. caudtus, B. calyciflorus, B. falcatus, B. ruben, Macrochaetus sericus, Lepadella bicornis, Cephalodella gibba* and *Filina longiseta*.

Reappearance of the aforesaid species from March onwards was mainly caused by favourable climatic conditions for their growth, reproduction and development Kumar (1994), Vijaykumar (1996), Vijaykumar *et al.*, (1999).

The higher number of rotifera was found during summer months. This may be due to higher concentration of nutrients like phosphate and ammonical nitrogen found suitable for the propagation of algal bloom in the Reservoir (Michael, 1969;

Chakrabarthy and Saha, 1993). The rainfall and water temperature have an influence the development of rotifera population (Goel *et al.*, 1988). Sinha (1992), Baruah *et al.*, (1993) reported that the zooplankton density was high in the summer season. This may be due to the favourable physico-chemical condition of water and hence, higher phytoplankton density during this season.

In the present study, we found that the rotifera population showed direct and significant correlation with water temperature, pH, dissolved oxygen and nutrients such as orthophosphate and ammonical nitrogen (P>0.01) establishing the determinative (as conditional and material factor, respectively) situation of the environment.

It also clears from the present investigation that *Rotatoria neptunia* exhibited much resistance to adverse conditions of the environment existing specially Chandrampalli reservoir. Thus, their numerical strength can be used as the most sensitive bioindicator of the water quality and trophic condition of the habit.

Table 20.1: Monthly variations in the physico-chemical parameters in Chandrampalli reservoir

Month/ Year	*Atmos. Temp. °C*	*Water Temp. °C*	*pH*	*DO mg/l*	*CO_2 mg/l*	*TA mg/l*	*Cl_2 mg/l*	*PO_4-P mg/l*	*NH_4-N mg/l*
1997									
June	25.6	20.1	8.6	4.3	+	140	34.60	0.12	0.27
July	29.7	24.3	9.1	3.1	0.8	125	63.90	0.14	0.67
August	28.9	24.9	8.4	2.5	8.8	170	62.48	0.12	0.61
September	25.3	21.3	8.6	3.0	6.6	117	59.64	0.16	0.61
October	27.8	22.8	7.9	2.5	+	165	59.64	0.17	0.29
November	29.5	24.5	8.8	3.1	+	170	58.22	0.11	0.68
December	29.7	25.9	8.0	3.5	1.1	185	48.28	0.11	0.67
1998									
January	29.9	24.9	8.3	5.9	1.1	155	58.22	0.12	0.61
February	29.5	23.5	8.0	8.0	+	185	48.28	0.13	0.70
March	31.7	25.0	8.2	4.4	1.1	175	34.60	0.14	0.69
April	32.0	26.7	8.1	4.3	2.2	135	34.60	0.12	0.39
May	33.2	27.4	8.2	4.0	1.1	120	48.28	0.11	0.31

Table 20.2: Monthly average of rotifera populations in Chandrampalli Reservoir (Organisms/l)

	1997							*1998*				
Species	*Jun*	*July*	*Aug*	*Sep*	*Oct*	*Nov*	*Dec*	*Jan*	*Feb*	*Mar*	*Apr*	*May*
Brachionus angularis	30	16	16	08	38	23	02	22	-	35	32	38
B. Caudatus	18	06	05	04	07	09	-	-	-	-	10	12
B. Calyciflorus	02	01	02	-	01	03	-	-	-	-	01	02
B. falcatus	01	01	01	01	-	01	01	-	-	01	-	02
B. rubens	32	30	30	28	22	29	34	-	32	25	39	40
Keratella tropica	31	36	30	32	39	26	35	-	15	34	36	38
K. cochlearis	05	08	04	01	02	05	09	01	-	01	03	12
Macrochaetus sericus	01	01	01	01	01	-	01	-	01	-	01	04
Lepadella bicornis	05	02	02	03	03	01	01	-	01	01	02	02
L. ovalis	03	01	01	02	04	01	02	01	-	-	01	01
Lecane (monostyla) bulla	01	01	01	01	01	01	01	02	01	01	01	01
Cephalodella gibba	05	04	04	02	02	04	03	-	-	04	09	06
Asplanchna pilina	36	30	22	18	38	10	10	-	17	39	28	38
Filina longiseta	03	02	01	01	01	01	01	-	-	-	02	01
Rotatoria neptunia	38	32	28	19	46	17	09	16	18	12	19	36

References

APHA, AWWA, WPCF (1985). Standard method for the examination of water and waste water. 16th ed., 1268.

Arora, H.C. (1966). Studies in Indian rotifera part-III, I. *Zool. Soc India*. 16, 1-6.

Baruah, A.K., D.K. Singh, A.K. Sinha and U.P. Sharma (1994). Diel variation of phytoplankton in tropical wet land of North Bihar. Environ. and Ecol. **12**(4): 30-32.

Bais, V., Vatheesh, S. and Agarwal, N.C. (1992). Benthic macroinvertebrates in relation to water and sediment chemistry. J. Fresh Water. Biol. **4**(3): 185-191.

Chakrabarthy, T.K. (1990). Cladoceran and copepoda fauna of Bhagalpur and their population dynamics. Fresh water Biol. **(2)**: 153-157.

Chakrabarthy, D. and Saha, G.K. (1993): The physico-chemical and Hydrobiological feature of three stagnant water sources in Darjeeling hills. Geobios. 20: 61-65.

Goel, P.K. Kulkarni, A.Y. and Khatwar, S.T. (1988). Species diversity in phytoplankton communities of few fresh water bodies in Sourth Western Maharashtra. Geobios. **15**: 150-156.

Jona, B.B. (1973). Seasonal periodicity of plankton in fresh water pond in West Bengal, Indian Int. Rev. Hydrobiol. **58**: 127-143.

Kumar Arvind (1994). Role of species diversity of aquatic insects in the assessment of population in wetlands of Santhal Parganas (Bihar). J. Envi and Pollution. **1(3 and 4)**: 117-120.

Michael, R.G. (1969). Seasonal trend in physico-chemical factor and plankton of fresh water fish and their role in fish culture. *Hydrobiologia*. **30**: 144-160.

Michael, R.G. (1973). Rotatorian : A guide to the fresh water organism. *J. Madurai Univ. Suppl.* 23-26.

Sharma, B.K. and Michael, R.G. (1980). Synopsis of Taxonomic studies on the Indian Rotatoria. Hydrobiologia. **73**: 129-236.

Nayar, C.K.G. (1968). Rotifera fauna of Rajasthan, Hydrobiologia. **31**: 168-185.

Nasar, S.A.K. (1975). Seasonal variations in the physico-chemical and biological properties of shallow pond. Japan J. Ecol. **24**: 255-259.

Nidham, J.H. and P.R. Nidham (1962). A guide to the study of fresh water biology. Holder Bay Inc. Sanfrancisco. 108.

Patil, C.S. and Goudar B.Y.M. (1989). Fresh water invertebrates of Dharwad (Karnataka state, India) Prasaranga, Karnataka University, Dharwad. 1-144.

Singh, R. (1991). Rotifera fauna of Jamalpur, Munger, Bihar. *J. Curr. Biol. Sci.* **8**: 45-48.

Tonapi, G.T. (1960). Fresh water animals of India. An ecological approach. Oxford and IBH, New Delhi, India.

Vijaykumar, K. and S. Basalingappa (1992). Effect of physico-chemical factors on the seasonal Abundance of cladocera in Devikoppa Tank (Dharwad District). Journal of the Karnataka University Science, Vol.XXXVI, 229-234.

Vijaykumar, K., Devendra Holkar and Kuldeep Kaur (1999). Limnological studies of Chandrampalli Reservoir. Freshwater Ecosystem of India ed. by K. Vijaykumar. Daya Publishing House, New Delhi. 59-108.

Vijaykumar (1996). Rotifer fauna of Devikoppa tank, Dharwad, Karnataka, India. 4th Chapter, Advances in Fish and wide life ecology and biology (ed) Kaul, B.L. Daya Publishing House,New Delhi. 187-192.

Welch, P.S. (1948). Limnology, New York. McGraw Hill. 537.

21

Variations in the Physico-Chemical Characteristics of Lentic Water Body of Gulbarga

K. Vijaykumar, C. Padmavathi, Virupanna, Ramesha I. and Halli Anjinappa
Environmental Biology Reserach Unit,
Department of Studies in Zoology,
Gulbarga University, Gulbarga - 585 106, Karnataka

Abstract

The present article deals with a investigation of a freshwater tank at Gulbarga for some physico-chemical parameters like, temperature, pH, transparency, total alkalinity, dissolved oxygen, chloride, hardness and free carbondioxide. Significant variations in these parameters were observed throughout the period of study and the observations suggests that the tank to be a highly polluted one calling upon urgent measures to be taken to prevent its further deterioration.

Introduction

It is a well-known fact that almost all-human activities modify the quality of our water resources. Whereas on a wider front, urbanization and industrialisation activity led to the pollution of water, in short- terms, direct human activities in the form of washing, bathing, offering of religious materials, and cattle wading may prove to much more effective. These factors are often found to be the culprit in pollution of small water bodies normally situated in the urban areas. The quality of water in this water body has deteroirated greatly resulting in their infestation by a variety of vegetation.

Present study has been carried out on a small water body called Jagath Tank, which is under the control of Parkithwar Fisherman's Co-operative Society of Gulbarga. This tank is being extensively used for fisheries activities. Investigations on several such water bodies have been carried out during past fifteen years in the region of Hyderabad-Karnataka area by Vijaykumar and Paul (1990 and 1994); Pappa (1994); Shivasharnappa (1996) and Paul and Verma (1999).

The study has been designed to understand the factors responsible for hampering the growth of fish and to find out the magnitude of pollution.

Material and Methods

The Study Area

Gulbarga is a township situated in the northern part of Karnataka State (76°-04′ to 77°-42′ N and 16°-12′ to 17°-46′), experiences semi-arid climatic conditions typical of most areas of Deccan Plateau. Jagath tank is a permanent standing water body situated in the heart of the city. This tank spreads over an area of 0.275 sq. km. The mean depth is 3.5 m meters during the monsoon and minimum being 1.2 m during summer season. The eastern, northern and southern boundaries are with roads while agricultural field's border a part of the western boundary. This tank is linked through a narrow channel towards the northern point with the moat of the Gulbarga fort, which is adjacent to the northern boundary, which received sewage from a part of the city, which ultimately reaches the tank. This tank is fully utilized by the Parkithwar Fisherman Co-Operative Society Ltd., Gulbarga for the fisheries activities since last 30 years.

Monthly samples were collected from April 1999 to March 2000. Observations were made for atmospheric temperature, water temperature, pH, dissolved oxygen, free carbondioxide, total alkalinity, chloride and amongst the nutrients orthophosphate and ammonical nitrogen as per methods described by Standard Methods (APHA, 1985) and Trivedy and Goel (1986).

Results and Discussion

Monthly variations in the physico-chemical parameters were presented in Table 21.1.

It is establishing fact that physico-chemical conditions of water determined the status of an aquatic eocsystem. These parameters also can be influenced greatly by the climatic conditions of the area (Vijaykumar and Paul, 1990). Atmospheric temperature was maximum (35°C) in the month of May and minimum (25°C) during January. The minimum water temperature was in the month of January, which is increased to a maximum value of 28°C during May and thus showing an annual fluctuations of 8°C.

The annual variation of pH was small and the maximum pH was noticed during March and lowest in the month of August, 1999. The pH remained between 7.8 to 8.9, which is rather more suitable for fish growth. A pH below 5 or above 11 is highly toxic to fish (Smitherman and Boyd, 1974). The pH greatly influences the toxicity of many other pollutants like ammonia and hydrogen sulphide.

The Dissolved Oxygen concentration indicated the gradual increase during post-monsoon season towards winter. The maximum (8.4 mg/l) dissolved oxygen concentration was recorded in the month of October and December 1999 and the lowest was recorded in the month of March. Low values of dissolved oxygen might be attributed to less phytoplankton population and increased utilization by organic decay (Khan and Khan, 1985, Singhal *et al.*, 1986).

The free carbondioxide concentration was highest in the month of August and lowest was recorded in the month of May and January, showed distinct peaks in the different seasons. The concentration of free carbondioxide is also crucial for the growth of fish, well aerated with low pollution usually have no or very little free carbondioxide. If the oxygen is in net saturation, a concentration of free carbondioxide upto 30 mg/l can be well tolerated by fish. Through the unit for free carbondioxide has been kept at 25 mg/l (Anonymous, 1968), generally its concentration should be 4 mg/l in ponds used for fish culture as the upper level oxygen cannot be maintained. In the presesnt study the free carbondioxide remains frequently at higher levels copuled with lower oxygen values indicating conditions of stress to fish. However, no correlation with pH values could be observed as also noticed by Ayyappan and Gupta (1981).

Table 21.1: Monthly average values of physico-chemical parameters of Jagath tank from April, 1999 to March, 2000

Parameters	*Months*											
	April	*May*	*June*	*July*	*Aug*	*Sep*	*Oct*	*Nov*	*Dec*	*Jan*	*Feb*	*March*
Atmospheric Temp. °C	33	35	30	30	29	25.8	29	30	25.5	25	29	32
Water Temp. °C	27	28	26	26	25	21	25	25	22	20	24	27
pH	8.6	8.5	8.8	7.8	8.0	8.3	8.6	8.2	8.5	8.7	8.4	8.9
Dissolved Oxygen (mg/l)	4.2	5.4	5.6	7.2	6.8	5.6	8.4	7.4	8.4	5.4	4.6	3.4
Free Carbondioxide (mg/l)	6.6	4.4	12.1	4.4	16.5	15.5	12.2	13.2	12.2	4.4	14.3	7.7
Total alkalinity (mg/l)	375	360	315	305	315	300	290	300	310	325	335	380
Chloride (mg/l)	224.36	238.36	71.10	38.34	52.54	45.44	220.1	238.56	247.08	224.36	224.36	235.70
Orthophosphate (mg/l)	1.8	2.0	2.2	1.4	1.6	1.8	2.0	1.4	1.7	1.8	2.2	3.0
Ammonical Nitrogen (mg/l)	0.8	0.7	0.9	0.6	0.5	0.7	0.9	0.6	0.7	0.8	1.1	1.4

Total alkalinity exhibited conditions varied with distinct seasonal trend, the increasing values towards April 1999 and March 2000. The high alkalinity values signify that this reservoir does not fall under the category of less productive water bodies (Phillipose, 1959).

The level of chloride fluctuated between 38.34 to 48.0 mg/l. The upper permissible chloride in both irrigation and drinking water is 600 mg/l (ISI, 1974). High concentration of chloride is often considered to be indication of pollution. In the present study the chloride concentration was high during all the months except during the rainy season.

The nutrients play an important role in metabolic activities of the plant and animals. Phosphorus is also one of the major nutrients responsible for biological productivity. It is usally found in the aquatic environment as biological trace of phosphorus. During the present investigation only inorganic phosphorus (orthophosphate) was estimated. Availability of inorganic phosphorus concentration is high during February and March, 2000.

Ammonical nitrogen content in water body fluctuated from 0.5 mg/l during rainy season to 1.4 mg/l during the summer season. Maximum concentration noticed in surface water during summer appears to be due to decaying organic matter provided by dense vegetation.

From the present study it may be concluded that the physico-chemical parameters of Jagath tank are indicating it as a highly polluted water body and the need for preventive measures to check its further deterioration.

References

APHA, AWWA, WPCF (1985). Standard methods for the examination of water and wastewater. American Public Health Association. Inc. New York 16th Edition. p. 1268.

Ayyappan, S. and Gupta, T.R.C. (1981). Limnology of the Ramasamudra tank : Hydrography, Mysore. J. Agri. Sci. 15: 305-312.

Khan, A.I. and Khan, A.A. (1985). Physico-chemical in Seika Jheel at Aligarh. Envi. and Ecol. 3: 269-274.

Pappa, A. (1994). Limnological studies of Gobbur tank. M.Phil Dissertation. Submitted to Gulbarga University, Gulbarga.

Paul Ravindra and Verma, S. (1999). Hydrobiological studies of tropical standing water body with special reference to nutrient status and primary production.

Phillipose, M.T. (1959). Freshwater phytoplankton of Inland fisheries. Proc. Symp. Algology. (ICAR). 272-291.

Smitherman, R. and Boyd (1974). Intensive management of water for fish production. In : M.C. Blount (ed): Water resource utilization and conservation in the environment fort valley state college, Fort Valley, Georgia. 451.

Shivasharanappa, J. (1996). Population dynamics of aquatic insects in freshwater bodies of Gulbarga. M.Phil. Dissertation submitted to Gulbarga University, Gulbarga.

Singhal, R.N., Swarn Jeet and Davis, R.W. (1986). The physico-chemical environment and the plankton of managed ponds in Haryana, India. Proc. Ind. Acad. Sci. (Anim). 95: 353-383.

Trivdoy, R.K. and Goel, P.K. (1986). Chemical and biological methods for water pollution studies. Enviornmental Publication, Karad (India). pp. 247.

Vijaykumar, K. and Paul Ravindra (1990). Physico-chemical studies of the Bhosga reservoir in Gulbarga, Karnataka, J. of Ecobiol. 2(4): 332-335.

22

Insect Abundance in Relation to Physico-Chemical Characteristics of Pond Water at Gulbarga, Karnataka

K. Vijaykumar, Ramesha, I., Halli Anjinappa, Virupanna and C. Padamavathi
Environmental Biology Research Unit,
Department of Zoology,
Gulbarga University, Gulbarga - 585 106, Karnataka

Abstract

Water quality on account of some physico-chemical parameters and aquatic insect population of sewage fed pond at Jagath tank during the period of November, 1998 to October, 1999 has been investigated. Insect abundance both in qualitative and quantitative was observed in different seasons with change in some of the physico-chemical parameters.

Introduction

Aquatic insects are a minor fraction of all insects, probably numbering no more than 3 to 5 per cent of all species. The limited number is probably that the result of the limited amount of freshwater habitat is comparison with the land surface. Yet these insects are taxonomically diverse, are fascinating structure and biology. Aquatic insects spend part or all of their lives in soil or water exhibit special structural and behavioural adaptations to the physico-chemical and biotic conditions found in each. These insects form an important link in the nutritional cycle of an aquatic ecosystem, as they constitute a bulk of food for fishes. Extensive work has been done on the seasonal hydrobiological conditions of freshwater bodies of India (Sinha and Sinha, 1990; Kaushik *et al.*, 1990, Pandey *et al.*, 1992; Singh, 1993; Arvindkumar, 1994) but information on the interrelationship between aquatic insect population and physico-chemical conditons is scanty. Hence, the influence of some physico-chemical factors on the abundance of aquatic insects of sewage fed tank at Gulbarga has been studied.

Material and Methods

Study Area

Gulbarga (76°-04′ to 77°-42′ N and 16°-12′ to 17°-46′) is situated in the northern part of Karnataka, experiences semi-arid climatic conditions typical of most areas of Deccan Plateau. Jagath tank is a permanent standing water body situated in the heart of the city. The eastern, northern and southern boundaries are with roads while agricultural field's border a part of the western boundary. This tank is linked through a narrow channel towards the northern point with the moat of the Gulbarga fort, which is adjacent to the northern boundary, which receives sewage from a part of the city, which ultimately reaches the tank.

Water samples of the tank were collected monthly from November 1998 to October 1999. Aquatic insects were collected with an insect collecting net made of nylon cloth having mesh size 40 to 80/cm^2 and were preserved in 5 per cent formalin and 90 per cent alcohol respectively. Physico-chemical analysis of water was done following the Standard Methods (APHA, 1985). Identification of insects was done based on the characteristics using Tonapi (1959); Michael (1973) and Gouder (1989).

Results and Discussion

During the present investigation, it was observed that the tank water was always greenish and alkaline. In monsoon seaons, the water of the tank was found to be moderately polluted, comparatively, due to dilution of water, in winter it was moderately polluted and in summer it was grossly polluted.

Survey of freshwater aquaculture tank at Gulbarga shows that 23 species of insects belonging to orders *viz.* Euphemeroptera, Odonata, Hemiptera, Coleoptera and Diptera occur throughout the year though their abundance varied depending upon climatic conditions. Two species of the insect density were exhibited during the course of the present study. First peak was observed in the April representing high magnitude of Dipterans and second peak was noticed during July and Hemipteran insects were high during all the seasons.

The transparency of Jagath tank was inversely proportional to the turbidity as observed by Sinha and Sinha (1990), while temperature is the most apparent factor which affects the seasonal cycle and abundance of insects in aquatic ecosystem (Kaushik *et al.*, 1990). Hardness and alkalinity were closely related to each other. During summer, alkalinity, hardness, chlorides and phosphate were recorded high and tank was more productive which supported rich fauna and flora. Sinha and Sinha (1990) have also noticed similar phenomena.

During winter seasons, when water was clear, *Cloeon* sp. was more in numbers on the submerged macrophytes, but in monsoon and summer when water turbid, the number was low.

Species of Odonata order flourished when macrophytic growth was luxuriant in tank during winter and summer season and has showed a significant positive correlation with chloride (r=0.52), phosphate (r=0.48), alkalinity (r=0.59), transparency (r=0.78) and total hardness (r=0.68).

Hemiptera group was recorded in high number during the summer and winter season. *D. rusticum* was noticed when the dense vegetation occurred and preferred the habitat with *R. elongata, Corixa sp., Laccotrephes ruber* attached themselves to the roots in water. *A. sardea, N. gluca, L. fluviorum, Lethocerus indicus, S. pectoralis* were present variably from good number to an abundance (Lachlain, 1977; Roy *et al.*, 1989 and Sinha and Sinha, 1990) have also reported that *Anisops sordea* in large numbers in moderately polluted waters. This group has significant positive correlation with total alkalinity (r=0.58), water temperature (r=0.34), chloride (r=0.45) and phosphate (r=0.59).

Dipterans like *Chironomus javanus, Tanypus bilobatus* was present in good number almost througout the year indicating tolerance too wide range of physico-chemical parameters. Pandey *et al.*, (1992) have also suggested that the presènce of these two species is an indication of extreme condition of tank pollution, where as the

Table 22.1: Showing abundance of insect population during different seasons of a year in relation to physico-chemical parameters of Jagath tank.

Season	*Physico-chemical parameters*		*Insect orders*				
			Euphemeroptera	Odonata	Hemiptera	Diptera	Coleoptera
Winter	Transparency	21-45 cm	+++	+++	+++	+++	+++
	Turbidity	25-89 cm	*Coleon sp.*	*Urothemis*	*R. elongata*	*Chironomus*	*C. lactobillis*
	Alkalinity	290-325 mg/l		*signata**	*D. rusticum*	*javanaus**	*H. olivaceous*
	Hardness	270-420 mg/l		*Enallagma sps.*	*A. sardea*	*Tanypus*	*Cybister limbatus*
	Chloride	140-145 mg/l		*Anax sp.*	*N. glauca*	*bilobatus*	*R. attenuta*
	Phosphate	1.4-2.0 mg/l			*Corixa sp.*	*Culex sp.*	*Halochares sp.*
	pH	8.0-8.5			*L. fluviorum*		*Bagous sp.*
					Laccotrephes		
					ruber		
					Lethocerus indicus		
					S. pectoralis		
Summer	Transparency	21-26 cm	+	+	++++	++++	++
	Turbidity	65-99 cm	*Caenis sp.*	*Urothemis*	*R. elongata*	*Chironomus*	R. attenuta
	Alkalinity	315-380 mg/l		*signata*	*D. rusticum*	*javanaus**	Bagous sp.*
	Hardness	340-350 mg/l		*Enallagma sps.**	*A. sardea*	*Tanypus*	
	Chloride	138-142 mg/l		*Anax sp.*	*N. glauca*	*bilobatus*	
	Phosphate	1.8-3.0 mg/l			*Laccotrephes*		
	pH	8.4-8.8			*ruber*		
					R. filiformis		
Monsoon	Transparency	15-18 cm	+	+	++	+++	+
	Turbidity	100-125 cm	*Cloeon sp.*	*Urothemis*	*R. elongata*	*Chironomus*	*Bagous sp.*
	Alkalinity	325-415 mg/l		*signata*	*D. rusticum*	*javanaus*	
	Hardness	270-360 mg/l			*Laccotrephes*	*Tanypus*	
	Chloride	60-81 mg/l			*ruber*	*bilobatus**	
	Phosphate	1.0-2.5 mg/l					

absence of *Chironomus sp.* along with the presence of many types of benthic species is an indication of good water qualities. This species population has significant correlation with total hardness (r=0.57), alkalinity (r=0.65), phosphate (r=0.38) and chloride (r=0.37). Among *Coleoptera, R. attenuata, C. latobillis* and *Halochares sp.* Were found on dead plant material as also observed by Kaushik *et al.* (1990). While these species were not available during summer season expect *Bagous sp.*

Aquatic insects as a whole in this tank have a significant positive correlation with total hardness, total alkalinity, chloride, phosphate and transparency.

References

APHA, AWWA, WPCF (1985). Standard methods for the examination of water and wastewater. American Public Health Association. Inc. New York, 16th Edn. 1268.

Arvind Kumar (1994). Role of species diversity of aquatic insects in the assessment of population in wetlands of Santhal Parganas (Bihar). J. Environment and Pollution 1(3 &4). 117-120.

Gouder, B.Y.M. (1989). Freshwater invertebrates of Dharwad, Karnataka, Published by Prasaranga Karnataka University, Dharwad.

Kaushik, S. Sharma, S.M.N. Saxena and Saksena, D.N. (1990). Abundance of Insects in relation to physico-chemical characteristics of pond water of Gwalior (M.P.). Proc. National Acad. Sci. India Vol. LX. Section B. Part II.

McLachlan, A. (1977). Composition, distribution, abundance and biomass of the macrofauna and microfauna of four Sandy beaches. Zoologica Africana 12: 279-306.

Michael, R.G. (1973). A guide to the freshwater organisms. J. Madurai Univ. Suppl. 7: 23-36.

Pandy, B.N., Jha, A.K. and Lal, R.N. (1992). Benthic macro-invertebrate communities as indicators of pollution in river Mahanand Katihar - Oikoassy. Vol. 9 No. 1 and 2. 35. 29.

Roy, S.P., Kumar, V., Pathak, M.S. and Sharma, U.P. (1989). Estimation of secondary productivity of Aquatic Insects in freshwater pond. J. Freshwater. Biol. 1: 109-112.

Singh, U.N. (1993). Studies on food and feeling behaviour of selected aquatic insects in artificial habitat. J. Comp. Physiol. Ecol., 1993. Vol. 18 No. 2, 69-71.

Sinha, K.K. and D.K. Sinha (1990). Seasonal variations in Biomass and production of Aquatic Insects in 9 Derlict Pond and a managed fish pond of Munger, Bihar. Envi. and Ecol. 8(4): 1231-1234.

Tonapi, G.T. (1959). Studies on the aquatic insects fauna of Poona (Aquatic - Heteroptera). Proc. Nat. Insti. Sci. India. 25(6): 321-332.

23

A Study of Bioindicators in Satluj River in the Region of Punjab

H. Kaur, J. Syal and S.S. Dhillon
Department of Zoology,
Punjabi University,
Patiala, Punjab

Abstract

River Staluj originates from Mansarowar lake in Tibet. It enters Punjab at Nangal, then flows through districts of Roper, Ludhiana, Jalandhar and finally joins the river Beas at Harike after which it leaves Indian territory close to Ferozepur to enter Pakistan. Due to its over-exploitation by industry and urban settlements for waste disposal, it has become highly polluted. Besides sewage and human ashes, it receives effluence from a variety of industrial units such as fertilizer manufacturing units, paper mills, hosiery, dyeing, electroplating etc. The physico-chemical and biological studies conducted for four years, at ten sites selected at various sections of the river along its course reveal that pollutants have altered the physico-chemical characteristics of water bringing about a series of changes in the biotic components. Water quality index calculated on the basis of physico-chemical characteristics at various sites reveal water to be unpolluted, slightly polluted, moderately polluted, excessively polluted and severly polluted. Four categories of organisms (planktons and invertebrates) have been indentified on the basis of their habitats. These are clean water species, moderately polluted water species, highly polluted water species and tolerant species.

Out of these four categories, clean water species and highly polluted water species have been labeled as bio-indicators since these can thrive only in clean and polluted waters respectively. Besides these categories, organisms found exclusively in waters receiving paper mill wastes, fertiliser factory wastes and sewage have also been identified and have been labeled as bio-indicators of a specific type of pollution.

Introduction

Aquatic organisms prefer variable habitats with defined physical, chemical and biological features. Variations in one or more of these defining factors can be stressful for some of the individuals and congenial for the others. The former may disappear completely from the affected areas, either due to death or migration while the latter continue to survive and flourish well. So the biological and physico-chemical studies, if super-imposed, could provide valuable information on potentialities of organisms as indicators of water quality.

River Satluj originates from Mansarowar lake in Tibet. It enters Punjab at Nangal, then flows through districts of Ropar, Ludhiana and Jalandhar, and finally joins the river Beas at Harike after which it leaves Indian territory close to Ferozepur to enter Pakistan. Due to its over-exploitation by industrial and urban settlements, water in the river is a grossly polluted. To assess water quality, this river was subjected to physico-chemical and biological analysis for a period of four years. Superimposition of data on these two types of studies yielded certain bioindicators which are described in this chapter.

Material and Methods

To assess water quality, ten sites were selected along the course of the river between Nangal and Harike reservoirs in such a manner that each selected site offered unique topographical features and a different set of pollution sources. The details of topography of the river and the selected sites have been described by Kaur *et al.*, (2000). Studies were conducted for four years, extending from May 1990 to April 1994.

Water samples were collected once in a month from all of the ten sites located along river Satluj and were subjected to physico-chemical analysis using methods suggested by APHA (1989). Water quality indices were calculated using nine chemical factors which include dissolved oxygen, bio-chemical oxygen demand, pH, total alkalinity, calcium, magnesium, sulphates, chlorides and nitrates.

For biological analysis, samples were collected with a wide mouthed polystyrene bottle measuring 250 ml. tied to a plankton net. Fifty litres of water was passed through the net to clllect one sample. Samples were preserved in 5 per cent formaldehyde solution at the sampling site and brought to the laboratory for microscopic examination in a Sedgewick-Rafter Cell. For identification of planktons and macro-invertebrates, books authored by Ward and Whipple (1959), Mellanby (1963) and Tonapi (1980) were consulted.

The physico-chemical and biological studies were superimposed to identify the bioindicator species of clean water, polluted water and some specific types of pollution.

Observations and Discussion

The water quality indices were calculated on the basis of nine physico-chemical factors. In all, three classes of water have been identified in the river. These are designated as class I (WQI between 80-100) representing clean to slightly polluted water, class II (WQI between 60-80) representing moderately polluted water and class III (WQI between 0-60) representing highly polluted water (Kaur *et al.*, 2000). For biological studies, samples collected from all the sites were examined to identify phytoplanktons, zooplanktons and macroinvertebrates. The distribution pattern of each species in relation to water quality was determined and categorization was done in the following manner:

Clean water species: Species that live only in waters having WQI values ranging between 80-100 (Class I). These have been placed in Group 'A'.

Moderately polluted water species: Species that live in waters having WQI values ranging between 60-100 (Class I and II). These have been grouped in Group 'B'.

Highly polluted water species: Species that live only in water with WQ1 between 0-60 (Class III). These have been placed in Group 'C'.

Tolerant species: Species that can live in waters having WQI values ranging between 0-100 (all types of water) and have been placed in Group 'D'.

On the basis of this categorization, the biota found in Satluj river at ten selected sites during a study of four years is enlisted as follows:

Clean Water Species

Phytoplankton

Chlorophyceae: *Hydrodictyon* sp., *Mougeotia* sp., *Ulothrix* sp., *Oedogonium* sp. and *Vaucheria* sp.

Cyanophyeae: *Lyngbaea* sp. and *Merismopedia* sp.

Bacillariophyceae: *Cocconeis* sp., *Gomphonema* sp. and *Synedra ulna*.

Zooplankton and Macro-invertebrates

Protozoa: *Chilomonas* sp.

Rotifera: *Ascomorpha* sp., *Brachionus forficula*, *Colurus* sp., *Cupleopagis vorax*, *Gastropus* sp., *Keratella serrulata*, *Lepocharis* sp., *Neptunis* sp., *Notops* sp., *Scaridium longicaudum*, *Trichocerca cylindrica* and *T. multicrinis*.

Nematoda: Nil

Tardigrada: Nil

Crustacea: *Ergasilus* sp., *Alonella* sp., *Bythotrephes* sp., *Moina brachiata* and *Sida crystallina*.

Insecta: Larvae of *Ephemera* sp., *Perla* sp. and *Helochares lividus*, and adults of *Polynema natans*.

Arachnida: Nil

Moderately Polluted Water Species

Phytoplankton

Chlorophyceae: *Characium* sp., *Stigeoclonium* sp., *Pediastrum simplex*.

Cyanophyeae: *Anabaera* sp., *Phoridrium* sp. and *Spirulina* sp.

Bacillariophyceae: *Cymbella aspera* and *Frustulia rhomboides*.

Zooplankton and Macro-invertebrates

Protozoa: *Oxytricha* sp. and *Euglypha ciliata*.

Rotifera: *Asplanchna* sp., *Dicranophorus forcipatus*, *Keratella valga*, *Lacane depresssa*, *L. luna*, *Lepadella* Sp., *Monostyla lunaris*, *Philodina* sp., *Pleurotrocha* sp., and *Trichotoria tetractis*.

Nematoda: Nil

Tardigrada: *Hypsibius* sp.

Crustacea: *Cypris* sp., *Canthocamptus* sp., *Eurycercus lamellatus*, *Limnocalanus* sp., and *Polyphenus pediculus*.

Insecta: *Tanypus* sp. (larvae), *Hydropsyche* sp. (larvae), *Corydalus* sp. (larvae), and *Nymphula* sp. (larvae),.

Arachnida: *Argyroneta aquatica*.

Highly Polluted Water Species

Phytoplankton

Chlorophyceae: Nil

Cyanophyeae: Nil

Bacillariophyceae: *Epithemia* sp., *Navicula cuspidata* and *N. naviculum*.

Zooplankton and Macro-invertebrates

Protozoa: *Sphenodiria* sp., *Paraquadrula* sp., *Chilophyra* sp. and *Phacus* sp.

Rotifera: *Keratella procurra*, *Mytilina* sp., *Proalinopsis caudatus* and *Rattulus* sp.

Nematoda: Nil

Tardigrada: Nil

Crustacea: *Moina dubia*.

Insecta: *Dicranota* sp. (larvae).

Arachnida: Nil

Tolerant Species

Phytoplankton

Chlorophyceae: *Cladophora* sp., *Closterium* sp., *Cosmarium* sp., *Scenedesmus* sp., *Spirogyra* sp., and *Tetraedron* sp.

Cyanophyeae: *Chroococcus* sp., *Micrococcus* sp. and *Oscillatoria* sp.

Bacillariophyceae: *Denticula* sp., *Fragilaria* sp., *Licromorpha* sp., *Melosira* sp., *Navicula ablonga*, *Pinnularia viridis*, *Synedra acus*, *Tabellaria* sp. and *Pleurosigma spenceri*.

Zooplankton and Macro-invertebrates

Protozoa: *Ceratium* sp., *Paramecium* sp., *Volvox* sp., *Difflugia* sp., *Euglena* sp., *Arcella* sp., *Centropyxis* sp. and *Oicomonas* sp.

Rotifera: *Anuraeopsis fissa, Brachionus angularis, B. bidentata, B. calyciflorus, B. plicatilis, B. quadridentatus, Cephallodella gibba, Colurella adriatica, Epiphanes* sp., *Euchalanis dilatata, Filinia longiseta, Harringia* sp., *Hexarthra* sp., *Keratella cochlearis, K. tropica, Lecane leontina, Monostyla bulla, Notholca* sp., *Notometta* sp. and *Rotifera* sp.

Nematoda: *Dorylaimus* sp., *Rhabditis* sp. and *Rhabdolaimus* sp.

Tardigrida: Nil

Crustacea: *Chirocephalus diaphanus, Cyclops* sp., *Mescocylops* sp., *Diaptomas* sp., *Nauplius* (larvae), *Alona* sp., *Bosmina longirostris, Chydorus* sp., *Ceriodaphnia* sp., *Daphnia* sp., and *Simocephalus expinosus.*

Insecta: *Isotoma palustris, Chironomus* sp. (larvae), *Culex* sp. (larvae), *Corixa* sp., *Chaborus* sp. (larvae), *Baetis* sp. (larvae), *Eristalis* sp. (larvae), *Dysticus* sp., *Tabanus* sp. (larvae), *Tipula* sp. (larvae) and *Hydrophilis* sp.

Arachnida: Nil

Out of these four categories, species belonging to Group A and C have been labeled as bioindicators of clean and highly polluted water respectively. Moderately polluted water species and tolerant species falling in groups B and D respectively, fail to indicate water quality as these can inhabit more than one type of water with pollution of variable levels. Gaufin and Tarzwell (1956), Hynes (1960) and Jolly and Chapman (1966) also laid down similar criteria for selection of bioindicators.

Besides these, some of the planktons were found exclusively in waters receiving paper mill wastes, fertilizer factory wastes and sewage. These organisms have also been labeled as bioindicators of specific type of pollution. These are:

Bio-indicators of paper mill wastes

Phytoplankton

Cyanophyeae: *Micrococcus* sp.

Bacillariophyceae: *Epithemia* sp. and *Navicula naviculum.*

Zooplankton

Rotifera: Notometta sp., *Dicranophorus forcipatus, Proalinopsis caudatus* and *Rattulus* sp.

Bio-indicators of fertilizer factory wastes

Phytoplankton

Cyanophyeae: *Phormidium* sp.

Bacillariophyceae: *Cymbella aspera.*

Bio-indicators of sewages

Zooplankton

Crustacea: *Limnocalanus* sp.

Acknowledgements

Authors are grateful to PSCST, Chandigarh and CSIR for the financial support.

References

APHA (1989). Standard methods for the examination of Water and Waste Water. American Public Health Assoc., New York.

Gaufin, A.R. and Tarzwell, C.M. (1956). Aquatic microinvertebrates communities as indicators of organic pollution in Lyttle Creek. *Sweage and Indust. Wastes.* **28**: 908-924.

Hynes, H.B.N. (1960). The Biology of Polluted Waters. Liwerpool Univ. Press.

Jolly, V.H. and Chapman, V.A. (1966). A preliminary biological study of the effects of pollution on farmer's creek and Cox's river, New South Wales. *Hydrobiologia.* **27**: 160-187.

Kaur, H., Syal, J. and Dhillon, S.S. (2000). Water quality index of the river Satluj, *Pollution Research.* Communicated.

Syal, J. (1997). Present ecological status of the river Satluj in the region of Punjab. Ph.D. Thesis, Punjabi University, Patiala.

24

River Water Pollution and Rural Health: A Review

B.B. Chanappagoudar and S.R. Salakinkop
Weed Control Scheme, Main Research Station,
University of Agricultural Sciences,
Dharwad - 580 005, Karnataka,

Rapid rate of industrialisation of urban section has accelerated water pollution and the areas adjacent to the industrial sites are affected due to dumping of waste material. Since industrialisation and population are complementary to each other. Problem of disposal of industrial waste has assumed paramount importance with setting up factories consuming large quantities of good quality water. The production of one tone of paper requires 250 tons of water, one ton of steel requires 150 tons and refining of one ton of oil requires 180 tons of water (FAO, 1977) (Table 24.1). These industries require large quantities of good quality water and discharge harmful effluents to natural streams. On the Hoogly river estuary alone there are nearly 159 industries of which 78 are jute, 12 textile mills, 7 tanneries and 5 formidable pulp and paper industry, 4 large distilleries and 53 miscellaneous industries. The Ganga river at Kanpur received the domestic water along with waste water from 45 tanneries, 10 textile mills, 3 woolen mills, 2 jute mills and a number of chemical and pharamaceutical industries.

Most rivers in Third world cities are literally large open sewers. Take the case of India, as documented in two volumes on *The state of India's Environment*: a Citizen's Report produced by a network of Indian NGOs coordinated by the Centre for Science and

Environment in Delhi. This points out that of India's 3119 towns and cities, only 209 have partial sewage and sewage treatment facilities and eight have full facilities. On the river Ganga alone, 114 cities each with 50,000 or more inhabitants dump untreated sewage into the river every day.

DDT factories, tanneries, paper and pulp mills, petrochemical and fertilizer complexes, rubber factories and a host of others use the river to get rid of their wastes. The Hooghly estuary is choked with the untreated industrial wastes from more than 150 major factories around Calcutta raw sewage pours into the river continuously from 361 outfalls.

Table 24.1: Water requirements for different industries

Industry	*Volume of water used*
Dairy	6-10 l/l milk
Sugar	1.5-4.1 l/kg cane crush
Distillery	20 l/l alcohol
Cotton textile	20-70 l/m cloth
Viscose rayon	1600 l/kg fibre
Pulp and paper	270-450 l/kg paper
Tannery	40-45 l/kg hide processed
Steel	20-50 l/kg steel
Coke-oven	1.5-2.0 l/kg
Urea	6.0-8.0 l/kg urea

River water polluted may be defined as "the addition to a natrual body of water of any material which diminishes its optimal economic use".

Industrial wastes the widest range of complex pollutants including various kinds of organic and inorganic chemicals found that pulp and paper mill wastes require oxygen for decomposition. Higher the amount of such wastes, the greater was the demand for oxygen from the receiving water. The polluting effects is due to high concentration of readily decomposable organic matter, causes deoxygenation and loss of aquatic life. Produce obnoxious odour of H_2S from SO_4 and increase the organic acid which lowers pH. The wastes may vary alkaline also. Industrial waste renovation and reuse technology has today become a major area of interest.

Table 24.2: Global data on waste water irrigation

Country and city	*Irrigated area (ha)*
Argentina, Mendoza	3700
Australia, Melbourne	10000
Bahrain, Tubli	800
Chile, Santiago	16000
China, all cities	1330000
Federal Republic of Germany, Braunschweig	3000
Other cities	25000
India, Calcutta	12500
All cities	73000
Israel, several cities	8800
Kuwait, several cities[a]	12000
Mexico, Mexico city	90000
All cities[a]	250000
Peru, Lima[a]	6800
Saudi Arabia, Riyadh	2850
South Africa, Johannesburg	1800
Sudan, Khartoum	2800
Tunisia, Tunis[a]	4450
Other cities[a]	2900
United States of America, Chandler, Arizona	2800
Bakersfield California	2250
Fresno, California	
	1625
Santa Rosa, California	1600
Lubbock, Texas	3000
Muskegon, Michigan	2200

[a] includes planned expansion of existing reuse

Industrial effluents in general and effluents of textile industry in particular in Rajasthan pose serious environmental problems both for human and animal health. Intenstive use of caustics acids, dyes and their copious discharge adversely effect the open lands and agricultural fields, pollute the rivers and undergrund water (Aggarwal and Praveen Kumar, 1990).

In terms of impact of human health in Third world cities, the dangers from most toxic industrial wastes are probably more localised and more open to swift and effective government control than those from other industrial pollutants. With regard to water pollution, there are usually four main soucres: sewages, industrial effluents, storm and urban run-off and agricultural run-off. Agricultural run-off is often an 'urban' problem since water sources from which an urban centre draws may be polluted with agricultural run-off and contain dangerous levels of toxic chemicals from fertilizer and biocides.

All Third World cities have much more serious "nonpoint" sources of water pollution than cities in the North because of the lack of sewers and the inadequacies in garbage collection services. A comprehensive sewage and storm drainage system makes it much easier to control water pollution since the wastes collected by this system can be treated, before being returned to rivers, lakes, estuaries or the sea. A comprehensive solid-waste collection and disposal system greatly limits the amount of solid waste that can be washed into water bodies. In the absence of such systems, much of the liquid wastes from households and businesses (and often industries) and a considerable proportion of the solid wastes end up washed into the nearby streams, rivers or lakes, greatly increasing the biochemical oxygen demand.

A shortage of water adds greatly to the problem of disposing of wastes especially liquid wastes from industries and sewage. Large volumes of water dilute wastes and can render them much less dangerous; in addition, bacteria in the water break down organic wastes, as long as the volume of wastes relative to the volume of water in not too great. Mexico City, like most of the largest cities and many smaller cities, is facing the mounting cost of increasing the water supply. Over-exploitation of underground water sources has made the city sink - in some areas by up to nine meters. New sources of water are at a considerable distance and these have to be pumped to 1000 m or more which adds considerable both to costs and to fuel consumption.

Mexico city is just one of many cities which developed in an area with limited water resources and which has now outgrown the capacity of the region to provide adequate, sustainable supplies. Dakar, the capital of Senegal, provides an example of a smaller city which is also facing serious problems of water supply. Water has to be drawn from ever more distant sources, as local groundwater supplies were fully used (and polluted) and local aquifers over-pumped, resulting in saltwater intrusion; a substantial part of the city's water has to be brought from the Lac de Guiers, 200 km away. Hundreds of urban centres which developed in relatively arid areas have grown beyond the point where adequate supplies can be tapped from local or even regional sources. Many of the coastal cities in Peru, Bangkok and Jakarata are two among many major coastal cities with serious problems of subsistence from drawing too much from underground aquifers and from saline intrusion into such ground waters, in Jakarata, many shops, houses and offices can no longer drink the water from the wells they use because of saline intrusion.

Most of water pollution can be grouped into three categories: liquid organic wastes; liquid inorganic wastes, and waterborne or water-based pathogens.

Liquid inorganic wastes

These are not broken down in water in the same way as organic wastes but for most, their dilution in large water bodies renders them harmless. Many such wastes kill animal and plant life, unless diluted sufficiently. Some inorganic wastes can become concentrated up the food chain to fish or through other fresh-or salt-water products to the point where they can kill or do severe damage to the health of humans who eat them. Wastes which include certain chemical elements known as heavy metals (Cd, Hg, Pb, Ni) or some of their componds can be particularly dangerous. Many of the pollution incidents which have resulted in the largest number of deaths and serious injuries from water pollution have arisen from human ingestion of fish or crops contaminated with heavy metals or their compounds.

Liquid organic wastes

These can be termed 'oxygen demanding' wastes since when disposed of into water, bacteria and other micro-organisms combine with oxygen dissolved in the water to break them down. The biochemical oxygen demand of such wastes is a measure of how much oxygen dissolved in the water they will need to be broken down and as such, is one of the most widely used indicators of pollution. Liquid organic wastes include sewage, many wastes from industries (especially industries processing agricultural products) and run-off from rains and storms which picks up organic wastes from land, before flowing into streams, rivers, lakes or seas. As the concentration of dissolved oxygen decreases, so fish and aquatic plant life suffer or die. Some portions of rivers or lakes which receive large volumes of organic wastes can have all their dissolved oxygen used up. They then lose their ability to break down these kinds of wastes and become black and foul smelling.

Waterborne or related pathogens

Many pathogens (disease causing agents including bacteria, viruses and worms) are spread in water—either through human ingestion of contaminated water or because water provides the habitat for intermediate hosts. Much the most common and widespread problem is pathogens from human excreta which contaminate water supplies. Typhoid, diarrhoeal diseases and cholera are among the diseases spread in this way. Contaminated water also has a central role in the transmission of many intestinal worms.

Water quality specifications

There are several basic characters used to evaluate waste water for its quality. They include BOD, Chemical Oxygen Demand (COD), the concentration of suspended solid particles, the pH of water and the quality and type of micro-organisms present (Anon, 1976).

According to specifications prescribed by Indian Standard Institute (ISI) (Table 24.3), the tolerance limits for industrial effluents discharged into public sewers is applicable to distilleries which are located in urban areas and are required to discharge effluents in public sewers (IS 3306). IS 3307 which envisages the tolerance limits for industrial effluents discharged on land for irrigation purposes is applicable for distilleries located in rural areas.

Characteristics Industrial Effluents

The sugar factory effluent, when disposed off in fresh water, renders submerged macrophytes like Ceratophyllum, Hydrilla, Nitella and Ottelia, unfit for growth and survival (David and Ray, 1966). Spent wash was a major pollutant because of its high organic content. The high Biological Oxygen Demand (BOD) caused depletion of dissolved oxygen and proved very harmful to aquatic life.

Table 24.3: ISI standards for discharge of industrial effluents

	Characteristics	*Tolerance limit for industrial effluents discharged*	
		Into inland surface waters IS 1974	*On land for irrigation IS 1977*
1	BOD (5 days at 20°C)	30	100
2	COD	250	-
3	pH	5.5-9.0	5.5-9.0
4	Total Suspended Solids	100	200
5	Total Dissolved Solids (inorganic)	2100	2100
6	Temperature (°C)	45	-
7	Oil and Greece	10	30
8	Phenolic compounds	1.0	-
9	Cyanides	0.2	-
10	Sulphide	2.0	-
11	Floride	2.0	-
12	Total residual chlorine	1.0	-
13	Insecticides	Absent	-
14	Arsenic	0.2	-
15	Cadmium	2.0	-
16	Cromium (hexavalent)	0.1	-
17	Copper	3.0	-
18	Lead	0.1	-
19	Mercury	0.01	-
20	Nickel	3.0	-
21	Selenium	0.05	-
22	Zinc	5.0	-
23	Chloride	-	800
24	Boron	-	2
25	Sulphate	-	1000
26	Per cent sodium	-	60
27	Ammonical nitrogen	5.0	-
28	Conductivity (μ mhos/cm)	-	2-1000

(All values except pH and temperature are in mg/l)

Pearson (1972) analyzed fruit proceessing waste waters for total salt content, ionic concentration and organic matter to assess their suitability for irrigation. He observed high COD value and excessively high concentration of sodium and chloride. The sodium and chloride ions in waste water were due to bring solution used in quality grading process.

Presence of organic acids, phenols, some inorganic and neutral compounds in spent liquors from coke oven section of steel plants are toxic to plants and animal life (Gupta and Mozumdar 1973). Balakrishna Rao *et al.*, (1977) have studied waste water from sugar industry, acetate factory textile mills and breweries for various parameters and reported that most of the effluents were acidic in nature and high in total solids, BOD and COD. A higher concentration of dissolved solids, ammonia, urea, sulphate, phosphate, flouride and organic with low concentration of BOD and COD were observed in the effluent from fertilizer factory producing ammonia, urea and phosphate group of fertilizers. Waste water of pharamaceutical company had high amount of total dissolved solids, sodium, chloride and high COD.

The waste water of tapioca starch industry was highly organic, containing low concentration of nitrogen and phosphorus and pH values ranging from 3.8 to 5.2. The cause of pH variation was attributed to addition of H_2SO_4 in the extraction process and release of some prussic acid by tapioca roots (Pescod and Thanh, 1977).

Survey of sugar factories in wastern Uttar Pradesh and analysis mixed sugar mill wastes of different factories showed high BOD (805 mg/l to 1660 mg/l) and COD (1011 mg/l to 2238 mg/l), respectively. Besides, good amount of oil and grease (16 mg/l to 54 mg/l) were also present (Verma *et al.*, 1978).

Effluent samples from paper factories were alkaline and contained considerable amount of nitrate nitrogen (313 to 368 mg/l). Seshsayee paper and Board mill effluent contained large amount of suspended and dissolved solids resulting in higher BOD and COD. Total phenol concentration was extremely higher in the effluent from Amaravathy. So also contents of carbonates, bicarbonates and sulphates were high in Amaravathy factory effluents whereas, calcium, magnesium and chloride concentrations were very high in Seshsayee factory effluent. They quoted that such type of variation in the constituents of effluents might be due to different types of raw materials employed as well as effluent treatment methods adopted by comparies (Rajannan and Oblisami, 1979).

Tannery effluent is highly coloured and foul smelling with electrical conductivity of 3.2 µ mhos/cm and pH 7.8. The suspended and dissolved solids (328 mg/l) also contained considerable amount of chlorides of sodium, magnesium and sulphate and magnesium which are known to be harmful for plant growth. The BOD and COD were high (752 mg/l and 4600 mg/l, respectively) (Kamalam and Raj, 1980). Effluent of Mysore Paper Mills, Bhadravati was light brown to dark brown in colour, bad smelling and contained high amount of suspended solids (933.6 mg/l) and dissolved solids (948.3 mg/l). BOD of this effluent was less than ISI limit prescribed for land irrigation. The pH ranged from 6.2 to 10.9 and electrical conductivity from 560 to 1400 us/cm. It was high in sodium (6.1 mg/l). The concentration of major nutrients was low (Srinivasamurthy, 1983).

Effect of different effluent on soil properties and crops

Effluent applied to soil undergoes a variety of physical, chemical and biological transformations. Some contaminants become fixed in soil, others are held on exchange complex where plants can utilize them, while others may leach to the water table below.

Working on pulp mill effluents found that when sodium absorption ratio of paper mill effluents exceeded 9.0, infiltration rate of soil was retarded. This was attributed to the sodium replacing calcium and magnesium from the exchange complex to soils resulting in the dispersion of soil aggregates (Blosser and Owens, 1964).

Narsimharao and Narasimharao (1992) indicated that the Bhadrachalam paper board industry effluent (Andhra Pradesh) could be safely used for irrigation to rice and cotton on alluvial soil having loamy to sandy loam texture, as these crops are tolerant to salinity and are able to give economic yields. But irrigation to tobacco and chilli with this effluent water led to poor quality produce and reduced crop yield.

Effluent discharged from Seshasayee paper and Board Ltd. (Pallipalyan) used for irrigation sugarcane crop on Typic Haplustalf for 15 years showed that pH, EC, organic carbon, CEC, exchangeable cations, available P, K and micronutrients and enzyme activities were more and there was practically no significant change in most of the soil properties (Palaniswami and Sreeramulu, 1994).

Fresh water lakes have Ca>Mg>Na>K : HCO_3>SO_4>Cl ionic composition. Analysis of water from different sampling stations of river Kali around Dandeli area, indicate an interesting relationship between the relative proportions in ionic composition, degree of population and eutrophication (Table 24.4). Upstream of the river at points which are less entrophic have the shifting of Mg and SO_4. The position of Na changes still further in case of the sampling station S_3, which in the present discussion is regarded as most polluted; it shows normal anionic series as suggested by Rodhe with large averages. But in case of sampling station S_4 the concentration of Na supersedes that of Ca, Mg and K and Cl shows larger average than SO_4. Therefore, this observation deserves consideration in the assessment of pollution and advancement of eutrophication.

Kali river flowing through the industrial township of Dandeli, receives effluents from the Dandeli Ferro-Alloy's Ltd., The Plastic Factory, Indian Plywood Manufacturing Co. Ltd., and also The West Coast Paper Mills, on the outskirts of Dandeli. Among the above industries, The West Coast Paper Mills effluents plays an important role in alteration of river water quality around Dandeli area. It discharges its effluents into the river.

Sreshail and Manjunathaiah (1994) studied the composition of Uger Sugar works effluents collected at different points and Krishna river water (Table 24.5) and also studied the dilution of Uger Sugar works effluents on yield of sugarcane, maize, wheat (Table 24.6, 24.7 and 24.8). They recorded that 1:10 dilution of effluent and river water better than other dilution.

Jhorar and Malik (1991) reported that germination percentage, rate of germination and dry matter yield recorded significant initial increase than decrease with increasing effluent BOD (Table 24.9) maximum and minimum for germination percentage and dry matter yield were 500, 750 and 16000 1250 mg/l BOD, respectively. The initial increase in germination and dry matter yield with increasing BOD of the Effluent may have been due to the specification effect of some of nutrient elements present in the effluent.

Table 24.4: Relative proportions of Na and Cl in the ionic composition indicating the degree of pollution due to discharge of effluent into river Kali

	Cautions	*Anions*	*Distance from Dandeli New Bridge, Km.*	*Source and Degree of Pollution*
Normal proportions	Ca>Mg>Na>K:	HCO_3>SO_4>Cl		According to Rodhe, 1949, Pollution Negligible
Station S_1 near Moulangi	Ca>Na>Mg>K:	HCO_3>Cl>SO_4	4.0	Practically no Pollution; free from entry of effluents
Stations S_2 near Dandeli New Bridge	Ca>Na>Mg>K:	HCO_3>Cl>SO_4	0.1	Less pollution; Because Ferro's Alloys Ltd., Plastic Factory and Plywood Factory do not significantly contribute towards water pollution
Station S_3 near Dandeleppa swamy temple	Na>Ca>K>Mg:	HCO_3>SO_4>Cl	2.0	High pollution due to entry of west coast paper mill effluents
Station S_4 near Kervada	Na>K>Ca>Mg:	HCO_3>Cl>SO_4	6.0	Low pollution; Water collects in the Bommanahalli pick up reservoir.

Table 24.5: Composition of Uger Sugar works effluents collection at different points and Krishna river water

Parameter	*Effluent before entering into field*	*Krishna river water*
Colour	Light brown	Clear
Odour	Tolerable	Odourless
Temperature (°C)	28-30	28-30
pH	7.04	7.80
Total Solids (mg/l)	1709	960
Total Dissolved Solids (mg/l)	840	662
Total Suspended Solids (mg/l)	764	298
BOD_{5d} (mg/l)	140	ND
COD (mg/l)	368	ND
Total nitrogen (mg/l)	7.6	Tr
Total phosphorus (mg/l)	11	Tr
Potassium (mg/l)	100	1
Sodium (mg/l)	68	55
Calcium (mg/l)	150	33.4
Magnesium (mg/l)	54	60
Carbonate (mg/l)	Nil	Nil
Bicarbonate (mg/l)	277	138
Chlorides (mg/l)	355	177.5
Sulphate (mg/l)	38.64	2.88
Iron (mg/l)	3.88	2.55
Manganese (mg/l)	0.80	0.40
Copper (mg/l)	0.22	0.16
Zinc (mg/l)	1.12	0.53

ND = Not detected
Tr = Traces

Sahal *et al.* (1983) used the effluent rich in Ca, Cl, HCO_3, total N and organic pollutants (BOD, 32 g/l). Germination percentage of rice seeds decreased with increase in the effluent concentration. Germination percentage seedling growth and speed of germiantion were retarded in higher effluent concentrations (25 per cent). The growth of seedlings was better in 5 per cent effluent than control. Further, it was seen that the root growth was more adversely affected than the shoot growth. The study, thus, reveals that the distillery effluent contains an excess of various forms of cations and anions which are injurious to plant growth. The concentration of these constituents should be reduced to beneficial level by diluting the effluent to 5 per cent which can be used for irrigation purpose as a substitute for chemical fertilizers.

Table 24.6: Effect of effluent dilution, amendment and their interactions on the cane and dry matter yields of sugarcane (t/ha)

Treatments	*Cane yield*		*Dry matter yield*	
11-1:10 effluent : river water	95.03		53.67	
12-1:25 effluent : river water	83.49		48.32	
13-1:50 effflunet : river water	77.19		44.73	
14-1:100 effluent : river water	72.77		42.09	
Mean	82.12		47.20	
	S.Em±	CD at 5 per cent	S.Em±	CD at 5 per cent
Between levels dilution (D)	0.915	4.119	0.353	1.589

Table 24.7: Effect of effluent dilution, amendment and their interactions on the grain and stover yields of maize (q/ha)

Treatment	*Grain yield*		*Stover yields*	
I_1-1:10 effluent : river water	80.17		84.51	
I_2-1:25 effluent : river water	72.99		80.02	
I_3-1:50 effluent : river water	70.28		79.04	
I_4-1:100 effluent : river water	69.23		76.58	
Mean	73.17		80.04	
	S.Em±	CD at 5 per cent	S.Em±	CD at 5 per cent
Between levels of dilution (D)	0.359	1.616	0.317	1.425

Table 24.8: Effect of effluent dilution, amendment and their interactions on the protein (per cent) in wheat and starch (per cent) in maize

Treatment	*Protein*	*Starch*
I_1-1:10 effluent : river water	13.17	64.40
I_2-1:25 effluent : river water	12.89	64.84
I_3-1:50 effluent : river water	12.84	64.95
I_4-1:100 effluent : river water	12.69	64.97
Mean	12.89	64.79

	S.Em±	CD at 5 per cent	S.Em±	CD at 5 per cent
Between levels of dilution (D)	0.049	0.220	0.072	0.325

Table 24.9: Germination percentage, rate of germination and dry matter yield of sorghum as affected by BOD of irrigation

Irrigation water (BOD mg/l)	*Germination per cent*	*Rate of germination/day*	*Dry matter yield (g/plot)*
0	90.0	28.6	15.6
250	93.3	25.2	16.7
500	96.7	25.7	17.4
750	93.3	24.4	20.4
1000	90.0	24.2	17.5
1250	79.7	19.2	0.0
2000	33.3	9.5	0.0
4000	26.7	7.1	0.0
8000	16.7	4.8	0.0
16000	0.0	0.0	0.0
32000	0.0	0.0	0.0
CD (5 per cent)	15.1	2.7	0.5

Application of distillery waste water upto 160 m^3/ha led to increase in dry matter production of mung bean and increase in both N and P uptake. Further increase in level of distillery waste water led to decrease in dry matter and N and P uptake. At 640 m^3/

ha application rate, the decline in dry matter was drastic and was even less than the control (Goyal *et al.*, 1995).

Asha and Dutta (1990) reported that irrigation of groundnut plants with raw distillery waste water resulted in a reduction of nodulation and an absence of fruit formation on the plant. However, when raw waste water was diluted to 50 per cent by tap water, the inhibitory effect was some what reduced. It is recognised that high levels of NH_4 - N and low pH are inhibitory to nodulation and nitrogen fixation. It was also noted that dilution of raw distillery waste water with stabilization pond effluent (1:1).

Impact of Sewage on Environment

Indiscriminate use of sewage and industrial effluents causes the development of "Sewage sickness" as a result of clogging of soil pores with decreased permeability. Lack of aeration sewage and industrial effenante sick soils cause anaerobic decomposition of sewage and industrial effluents which produces certain gases such as, methane, hydrogen sulphide and mercaptans etc. and these were found to create unhygienic conditions. High salt concentrations, heavy metals content (Cd, Pb, Hg, Ni, Cr, As, etc.), pathogenic organisms, pesticides and some organic compounds present in the effluents are major concern to utilize the effluents on crop land. Possible contamination of crop produce with human pathogen viz., *E. coli*, *Salmonella sp.*, *Mycobacterium* spp., *Enatmoeba*, polio virus and hepatitus virus have been widely reported.

Use of sewage in agriculture

The growing volume of municipal sewage waters, resulting from the rapidly increasing world's urban population, creates severe environmental and health hazards, such as uncontrolled flows of sewage polluting surface and ground waters, soil and agricultural products. Soil is not only a medium for plant growth but also a sink for waste solids and liquids. Different technologies have been developed to prevent this type of environmental degradation of which the use of land as a living filter for removing pollutants from waste water, especially is popular. This technology is viable, environmentally sound and cost effective.

Irrigation with sewage effluent provides crops with water and nutrients. At the same time it provides a convenient means of sewage disposal through land treatment, preventing potential health and environmental hazards caused by uncontrolled flow of waste water.

Under arid and semi-arid environmental conditions, irrigation is a pre-requisite for agricultural development, the importance of effluent as a source of irrigation water is as great as the need of its disposal.

However, there are some limitations imposed by sewage water due to pathogenic micro-organisms, high salinity and sodality, imbalanced nutrient content, organic pesticides and suspended solids. Under specific circumstances, presence of toxic levels of micro-elements and heavy metals make it unfit for irrigation.

The organic farming/sustainable agricultural production gives more emphasis to research and use of new organic sources of nutrients like crop residues, industrial wastes, sewage sludge and wastewater. Little work is being conducted on sewage use as both the source of irrigation and plant nutrients at the sponsoring institution and elsewhere. The available information is given below.

Potential source of plant nutrients

All the extractable nutrients (P, K, S, Ca, Fe, Mn and B) in sewage irrigated soils are higher than the soils irrigated with pump water. The heavy texture sewage treated soils will accumulate higher organic matter and nutrients than coarse textured soils. Total P and K content of sewage amended soils were 47 and 34 per cent respectively, then those of adjacent tubewell water irrigated soils.

Field experiment conducted by Salakinkop (1999) showed that sewage irrigated land recorded higher (about 23 per cent) grain (43.7 q ha^{-1}) and straw (54.6 q ha^{-1}) yield than borewell irrigated land. It also recorded higher dry matter and yield attributes (Table 24.10). Sources of irrigation differed significantly producing higher grain yield is sewage irrigation (41.9 q ha^{-1}) and alternate irrigation (39.5 q ha^{-1}) compared to borewell irrigation (35.5 q ha^{-1}). Straw yield also followed the same trend recording 57.7, 50.6 and 46.8 q ha^{-1} in sewage, alternate and borewell water irrigation, respectively. The grain yield was influenced mainly by yield components *viz.*, ears m^{-2}, ear weight, seeds ear^{-1} and test weight which were higher under sewage and alternate irrigations.

Among the fertilizer levels, 100 per cent RDF (45.1 q ha^{-1}) and 75 per cent RDF (43.0 q ha^{-1}) were on par with each other. This emphasized the importance of sewage irrigation in saving 25 per cent cost of fertilizer. Application of 75 per cent RDF and 100 per cent RDF to sewage irrigated land under sewage irrigation resulted in higher but similar grain and straw yield over other interactions. Same trend was noticed in both the lands wherever sewage and alternate irrigations were provided indicating saving of 25 per cent RDF with use of sewage water. Soil under sewage irrigated land contained significantly higher N, P, K, S and micronutrients (Cu, Fe, Mn and Zn).

There are many reports from all over the world that crop yields are significantly increased by irrigation with sewage water. In India, long-term field experiments at the National Envrionmental Engineering Research Institute in Nagpur have shown that medium intensity irrigation with waste water produced higher yields than irrigation with freshwater supplemented with standard dose of nitrogen, phosphorus and potassium (NPK), as shown in the table below.

Irrigation water	*Crop yields (tonnes per hectare per year)*				
	Wheat	*Moong beans*	*Rice*	*Potato*	*Cotton*
	(8)	*(5)*	*(7)*	*(4)*	*(3)*
Raw wastewater	3.34	0.90	2.97	23.11	2.56
Settled wastewater	3.45	0.87	2.94	20.78	2.30
Stabilization pond effluent	3.45	0.78	2.98	22.31	2.41
Fresh water +NPK	2.70	0.72	2.03	17.16	1.70

( ) = years of experimentation

Effect on soil properties

The pH of soils amended with sewage water (pH 7.9-8.3) was lower than that of soils (pH 8.6 to 9.0) irrigated with tubewell water. Organic matter, $CaCo_3$ and clay content were higher in sewage irrigated soils than tubewell irrigated soils.

A significant decrease in electrical conductance, pH and organic carbon has been observed in soils irrigated with sewage plus tubewell water.

Two year field experiment by Salakinkop (1999) indicated improvement in Soil maximum water holding capacity, BD and content of toxic heavy metals in soil did not increased significantly. Microbial population and their activity interms of dehydrogenase, phosphates increased due to sewage application.

Quality characters of crop produce

Potato tuber grown in field receiving irrigation showed high amount of protein, total nitrogen, total P and soluble N; the tubers had low amount of non-reducing and reducing sugars. This lead to contamination. Their concentration can be reduced by algal growth enhanced by sewage water and algae act as biofertiliser. However, some of microbes activity in soil is enhanced by sewage irrigation. N-fixers like Azotobacter and Azospitillum and BGA activity increases. Salakinkop (1999) showed that quality of crops interms of protein and gluten improved due to sewage water application (Table 24.11) and there was no accumulation of heavy metals (Cr, Ni, Pb and Cd) in plant parts above the threshold limit.

References

Anonymous (1977). Water-life blood of agriculture, Bhagirath. 24: 51-55. Pub. FAO.

Balakrishna Rao, K., Mithyantha, M.S. and Perur, N.G. (1977). The nature of some industrial effluent waters of Karnataka. Curt. Res. 6: 51-52.

Goyal, Sneh Chander K. and Kapoor, K.K. (1995). Effect of distillery waste water application of soil microbiology properties and plant growth. MKK Publication ISSN.

Gupta, G.C. and Mozumdar, A.K. (1973). Treatment of Ooke oven effluents. IAWPC commemmgration, pp. 154-165.

Guruswamy, R. (1988). Pollution control Hand Book. pp. 202-203.

Jhora, B.S. and Malik, R.S. (1991). Effect of an acidic organic industrial effluent on dissolution of calcite on germination and dry matter yield of sorghum. Annals of Arid zone, 30(2): 111-117.

Kamalam, N. and Raj, K. Effect of tannery effluent on germination and nutrient uptake of ragi. Madras Agril. J. 67(7): 441.

Pearson, C.A. (1972). Suitability of fruit processing waste water for irrigation. J. Environ. Qual. 1: 394-396.

Pescod, M.B. and Thanh, N.C. (1977). Treatment alternatives for waste water from tapicoa starch industry. Prog. Wat. Tech. 9: 563-574.

Pujar, S.S. (1984). Effect of distillery effluent irrigation on growth yield and quality of crops in vertisols. M. Sc. (Agril.) Thesis submitted to UAS, Dharwad.

Sahai, R. Jabeen and Saxena, P.K. (1963). Effect of distillery effluents on the seed germination, growth, pigment content and biomass on *Phaseolus radiatus* L. Indian J. Ecol. 14(1): 21-25.

Salakinkop, S.R. (1999). Conjunctive use of sewage and borewell water on performance of wheat (*Tritioum aestivum L.*) Ph.D. thesis submitted to UAS, Dharwad.

Srinivasamurthy, C.A. (1983). Studies on the effect of industrial effluents on soil properties and crop growth. Ph.D. thesis, UAS, Bangalore.

Rajannan, G. and Oblisami, G. (1979). Effect of paper factory effluents on soil and crop plants. Indian J. Environ. Hlth., 21: 120-130.

Verma, S.R., Bahel, D.K., Pal, N. and Dalela, R.C. (1970). Studies on sugar factories and their wastes on western U.P. Indian J. Environ. Hlth., 20: 205-218.

25

Histopathological Manifestations of Strigeoid Infection in *Heteropneustes Fossilis* (Bloch)

M.M.R. Nomani
Department of Zoology,
M.L.S.M. College,
Dharbhanga - 846 004

N.K. Dubey
University Department of Zoology,
L.N. Mithila University,
Dharbhanga - 846 004

Abstract

Tissue migrating as well as encysted strigeoid metacercariae, particuarly of *Diplostomum* sps; cause lesions in the tissues, like skin, liver and eyes in *H. fossilis*. Rapid invasion by large no. of cercariae produce haemorrhage, hyperemia, patechiae and necrosis in the superficial areas of body musculature and skin. In the eyes, some of the histopathological manifestations observed, were disorientation of the photoreceptor cells and diffused pigment epithelial cells. Displacement of retina from normal curvature and the location of free metacercariae in the peripheral retina probably cause milky appearance of the eyes, impaired vision and often blindness. Manifestation of "worm cataract" in some, is also not ruled out. Affected and invaded liver, revealed pushed aside tissue elements to make room for the parasites, engorged sinusoids and blood vessels and evidence of destructive and mechanical injury. It is hoped that the present work will give impetus to the application of histology as an adjunct to the diagonosis of fish disease.

Introduction

The Air Breathing fishes whcih thrive in plenty in the swamps and derelict water bodies of Mithilanchal, constitute one of the most economically important fishery resources of the region.

These fishes are often subjected to various parastitic diseases. Unsuitable food, bad treatment, lack of oxygen, temperature extremes or other adverse conditions make them weak and prone to various parasitic elements.

Among the maladies that afflict fish are some strigeoid borne diseases which are yet to be investigated from stand point of histopathology. Strigeoids are the members of the super family strigeoidea which have fish as their second intermediate hosts.

Despite the programmes launched to intensify development of inland fisheries, much remains to be known about the overall effects of worm burdens on fish biology and productivity.

The deleterious effects of tissue migrating phases of strigeoids and their metacercariae are often associated with extensive damage to vital organs. But unfortunately, this aspect of fish pathology continue to be unexplored.

Keeping in view the facts mentioned above, the present work was undertaken to record the histopathological manifestations of the infections of Diplostomum sps. in the fish host, *Heteropneustes fossilis*.

Material and Methods

The skin, muscles, liver and eyes were examined in saline solution under the dissecting microscope. The affected parts of the tissues were cut into pieces. They were fixed in the formal-acetic acid-alcohol mixture *i.e.* formaldehyde absolute alcohol, glacial acetic acid and distilled water in the ratio of 30:20:30 respectively. The fixed normal and infected tissues were further processed using standard histological techniques. Paraffin sections were cut at 5-7 microns and stained with H and E.

Results and Discussion

Histopathological examination of skin revealed haemorrhage, hyperemia petechiae and necrosis in the superficial areas of body musculature and skin. The tissue element appeared pushed aside to make room for Diplostomum sps metacercariae.

It appeared that the organisms could generally be considered as being parasite of little consequence as far as tissue reaction and inflammation were concerned. However, frequent evidence of marked destrcutive and mechanical injury was noticed.

Infected eyes were whitish and opaque, parasites were seen in the peripheral retina. Often the photoreceptor cell layer and the pigment epithelium were seen separated. Parasites occupied the cavity made by the separation of these layers. The photoreceptor cells were disoriented. Pigment epithelium cells were diffused and the retina was displaced from the normal curvature. The histopatholocial examination of the invaded liver tissue revealed that the hepatocytes had undergone degeneration. The normal configuration of hepatic cords was lost and the sinusoids and blood vessels were engorged. Hoffman (1975) has reported that a large number of ceracariae penetrate the skin of fish and once they are in the tissues, the cercariae bend upon itself and secrete a cyst wall around it. Lesions of the eye are numerous and varied in their etiology, but literature dealing with the lesions of the eyes due to metacercarial infection are not much. Some notable works have been contributed by Sato *et al.*, (1975) Lester and Huizinga (1977) and Dubey *et al.*, (1981). The liver of teleost does not show the diversity of pathology seen in the higher animals but liver is susceptible to a number of toxic and metabolic disturbances. In the present investigation the metacercariae of Diplostomum sps. present in the liver were not of the cyst forming type rather they all were actively moving and feeding on the liver tissues. The histopathological changes like necrosis and other mechanical damages were of much pronounced nature. The extent of host cellular reaction is dependent on the number of invasive elements. Host and parasite species differences may also be an important factor in regard to the damage done to the tissues. It has been observed that strigeoids produce leucocytosis, haemorrhage, hyperemia and necrosis. But once, the metacercariae get encysted, they tend to produce less damage to the host due to connective tissue encapsulation. It appears that the histological lesions observed in the present case are due to migrating nature of strigeoid larvae, possibly they migrate in search of the most favoured site. It was interesting to note that in spite of the considerable lesions, the fish appeared normal with no physical symptoms of any abnormal behaviour. The extent of damage caused by these parasites to wild fish stocks or hatchery stocks in India is unknown. But there are reports of significant growth reduction in the affected fishes owing to such infections. However, Ganapati *et al.*, (1954) and Chauhan *et al.*, (1958) have reported mortality in heavily infected fingerling stage of fishes. The mortality may be due to movement of the parasite within the tissues of the host or it may be due to toxins liberated by the parasites (Das and Das 1997).

The deleterious effects of strigeoids on fish production could be assessed from accurate histopathological examination of the invaded regions. Only then, a clearer picture of pathogenesis of strigeoid infection in fish is expected to emerge. The present work is just an attempt in this direction.

References

Chauhan, B.S. and Ramkrishna, G. (1958). On the occurrence of fish mortality due to helminthic infestation in a stocking tank at Nagpur (India). Indian J. Helminth, 10(1): 53-55.

Das, M.K. and Das, R.K. (1997). Black spot disease pp. 62-65. In "Fish and Prawn Diseases in India, Diagonosis and control", IFSI, Barackpore.

Dubey, N.K., Dubey, Usha and Pandey, P.K. (1981). Worm cateract of Heteropneusts fossilis (Bloch.), Curr. Sc. Vol. 50, No. 14, 651.

Ganpati, P.N. and Hanumantha Rao, K. (1954). On Black grub disease in the fresh water Carp Catla catla. Curr. Sc. 23: 401-2.

Hoffman, G.L. (1975). Lesions due to internal helminths of fresh water fishes, pp. 151-188. In "The, Pathology of Fishes". ed. Ribelin and Migaki. The Univ. Wisconsin press.

Lester, R.J. G. and Huizinga, H.W. (1977). Diplostomum adamsi Sp. n. description, life - cycle and pathogeneis in the retina of *Perca flavescens*. Can. J. Zool. Vol. 55 NO. 1, pp. 64-73.

Sato, T., Hoshina, T. and Horiuchi, M. (1975). On Worm cataract of Rainbow trent in Japan. Bull. Jap. Soc. Sci. Fisheries; 42(2): 249.

26

Acute Toxicity of Organo Phosphorous and Organo Chlorine Pesticides to the Fresh Water Climbing Perch, *Anabas Testudineus*

M.M. Jha and R. Kumar
Department of Zoology, M.L.S.M. College,
Darbhanga - 846004

Abstract

The toxic effect of organo phophorous (OP) and organo chlorine (OC) pesticides on *Anabas testudineus* shows several behavioural responses. When the fish were exposed to Dimethoate and Carbaryl for a period of 96 hours, the LC_{50} values for 24, 48, 72 and 96 hours were 8.43, 7.25, 6.25 and 5.25 ppm and 10.25, 9.25, 8.64 and 7.75 ppm respectively with regression equation while with logistic response model the values are 8.11, 7.58, 6.38 and 5.11 ppm and 10.0, 9.07, 8.69 and 7.62 ppm respectively.

The test fish exposed to different concentration of two pesticides exhibited more or less similar behavioural responses which were more pronounced in case of carbaryl exposure. The chief behavioural responses were faster and erratic opercular activity, surfacing and gulping of air, difficulty in respiration, sluggishness, profuse mucus secretion etc., all these bahavioural changes are reflections of signs of distress. The toxic potential of a particular chemical reduces the fitness of a population. It was thus felt necessary to have an accurate method for finding out toxic concentration of a given chemical to a given animal in a specific time. Acute toxicity test in term of LC_{50} are the most widely practised and accepted method for the determination of such toxic level.

Introduction

Industrialisation, urbanisation and increased population in India have modified environment beyond limits. The problem of environment pollution is more severe in developing countries like India, because of limited economic and technological resources than the developed countries. The disposal of industrial wastes posses a serious problem at present due to their diverse composition. Interaction of these components further increases their complexity. Industrial wastes may contain many toxicants such as acids, alkali, metallic ions, phenols, cyanide, pesticidies and many other organic and inorganic substances which have become a pervasive threat to the natural ecosystem.

Environmental contaminants have toxic effects on different types of organisms and affect biological processes at cellalar, population, community and ecosystem levels of organizations. The problem of toxicants affecting aquatic ecosystem is complex and there is continuing need for monitoring and mitigating the effect. A study of the structure and functions of a biological community in the waste receiving water and laboratory bioassays for determinging toxicants are some of the basic approaches that is required for.

Pollution Monitoring

Pesticidies have become an important food of modern agriculture to protect standing corps, stored gains and human belongings from pests and help preventing certain diseases. The average consumption of pesticides has increased sharply from 200 tons in 1955, 75,000 tons in 1985 and reached upto 1,00,000 tons in 1990 and is expected to be 200 thousand tons in 2000 A.D. (Mehrotra, 1990).

Pesticides after application eventually reach to the aquatic system is considerable amount *via* agricultural run off from land, contaminated ground water, bottom sediments etc., such contamination deterioates the water quality and affect the target as well as several non-target species also. These pesticides make the life of non-target species hell and amongst them fishes are the worst victims. Suffering of fishes to pesticide may be expressed either by their mortality or by alteration in their growth, development, reproduction, bio-chemistry and physiology, depending upon several factors like physical and chemical factors, concentration of pesticides duration and kind of exposure (Acute or chronic) and ability of fish to metabolize the pesticide absorbed. The toxic potential of a particular chemical reduces the fitness of a population. It was, thus, felt necessary to have an accurate method for finding out toxic concentration of a given chemical to a given animal in a specific time. Acute toxicity in term of LC_{50} are the most widely practical and accepted methods for determination of such toxic levels. Organophophorous and organochlorine pesticides are used by the farmers to cotrol the crop pests and diseases for higher agricultural productivity. However, it has led to a lone but steadily deterioration of aquatic ecosystem. These pollutants also destroy the quality of aquatic ecosystem and rander it unfit for various aquatic organism, particularly fishes (Sprague, 1969; Annes, 1975; Konar, 1981; Shafi and Alam, 1990; Sadhu, 1993; Borah and Yadav, 1995; Choudhary, 1996; Singh, 1999).

The acute toxicity of a few organo phosphate (op) compounds was less for air-breathing fish owing to the reduced intake of the toxicant *via* gills (Jacob *et al.*, 1982). While organochlorine (oc) shows high acute toxicity as reported by Johnson *et al.*, 1980 with carbaryl, (2 to 39 mg/l to invertebrates). In general oc compounds are more toxic than op compounds to fish. (Swarup *et al.*, 1981, Choudhary, 1996 and Singh, 1999).

The present study has therefore, been undertaken to establish the acute toxicity in terms of LC_{50} of pesticides to fish in relation to the behavioural responses, various methods like probits, graphic regression equation and logistic response are often employed in most of the bio-assay experiments for the routine analysis of LC_{50}.

Materials and Methods

Selection, Procurement and Maintenance of Test Fish

The adult living specimens of *Anabas testudineus* were procured from the local fish market and were brought to the laboratory in wide mouthed large earthen pots, half filled with natural water. The fish was washed with 0.1 per cent $KMnO_4$ solution to remove dermal infections, if any. Healthy fish of average length (9-13 cm) and weight (21-25 gram) were selected and acclimated in 40 litre glass aquaria to the laboratory conditions for a fortnight. Running tap water was used in all the experiments. A batch of 10 fish was transferred to the control and experimental tanks. They were fed with chopped goat liver every day *ad libitun* at 9.30 am. The feeding was stopped a day before the enitiation of the experiment (staice acute bioassay) and no food was supplied during the period of this test. However, during long-term exposure they were given food.

Analysis of Physico-Chemical characteristics of water

Physico-chemical characteristic of water were analysed following the method of APHA, AWWA and WPCF (1985). The pH was recorded by an electronic pH meter (Hanamake, Singapore).

Static Acute Bio Assay

Bioassays were conducted to determine the acute toxicity of the two test pesticides, Dimethoate and carbaryl, in term of LC_{50} values (concentration lethal to 50 per cent of the test fish) for 24, 48, 72 and 96 hrs following the methods of APHA, AWWA and WPCF (1985). Experiments were performed in twenty-litres rectangular glass aquaria and a group of ten fish were uses at a time for testing the toxicities of the pesticides. A series of six concentration of each pesticide in addition to control were used in order to make a fairly precise estimation of LC_{50} values. The running tap water was used as diluent for the test concentrations. The number of aquaria in each set was, thus, seven (one control +6 experimental).

The fish were kept in diluent water for atleast one day before the start of a test. Exposure units were static and not aerated during the 96 hours test. The pesticides were in logrithimic series. The healthy and well acclimatized fish were transferred one by one with the help of small hand net from the experimental ones.

The stock solution of pesticides were prepared by simple dilution technique given in standard methods APHA, AWWA and WPCF (1985). The fish were exposed to different concentration of Dimethoate and Carbaryl at intervals of 24, 48, 72 and 96 hrs were recorded as a measure of acute toxicity. The LC_{50} values for these time intervals were worked out statistically by logistic response modal (Byron and Brown, 1967) and Regression equation (Downie and Heath, 1970).

Results

The physico-chemical charateristics of the test water are enlisted in Table 26.1. The ratio of two individual pesticides on the basis of mortality denoted that carbaryl was more toxic than the Dimethoate to the fish *Anabas testudineus* (Carbaryl > Dimethoate).

The relative toxicity of the two pesticides obtained under static system have been shown in Table 26.2 and figure 26.1 - 26.10. The 24 hr LC_{50} value for Dimethoate and Carbaryl by Regression equation in Table 26.2 is 8.43 ppm and 10.25 ppm respectively while by logistic Response model the value is 8.11 ppm and 10.0 ppm in Table 26.5. The 48, 72, 96 hrs LC_{50} values for Dimethoate by Regression equation is 7.25, 5.25 ppm in Table 26.2, figure 26.9(A), while in Table 26.5 with logistic response model the values are 7.58, 6.38, 5.11 ppm Table 26.5, Fig. 26.9(B). The 48, 72, 96 hrs LC_{50} values for Carbaryl by Regression equation is 9.25, 8.64, 7.75 ppm, Table 26.2, Fig. 26.10(A) while it is 9.07, 8.69, 7.62 ppm with logistic response model Table 26.5, Fig. 26.10(B).

Table 26.1: Physico-Chemical Characteristics of the Test Water

Characteristics	*Unit*	*No. Of observations*	*Mean*	*Range*
pH		10	7.6	7.2 - 7.9
Temperature	0°C	10	24.6	23 - 26
Total Solids	mg/l	10	38.5	32.6 - 38.2
Dissolved Solids	mg/l	10	12.3	10.5 - 12.6
Dissolved Oxygen (DO)	mg/l	10	6.5	5.4 - 7.6
Free CO_2	mg/l	5	2.1	1.3 - 3.0
Total hardness	mg/l as $CaCO_3$	5	180.0	165.0 - 200.0
Total alkalinity	mg/l as $CaCO_3$	5	150.0	136.0 - 165.0

Table 26.2: LC_{50} values of the pesticidies with their Regression equations to *Anabas Testudineus*

	Pesticides	*Time (hr)*	*LC_{50} (mg/l)*	*Regression Equation*
1	Dimethoate	24	8.43	- 1300 + 160 x
		48	7.25	- 240 + 40 x
		72	6.25	- 200 + 40 x
		96	5.25	- 160 + 40 x
2	Carbaryl	24	10.25	- 3600 + 40 x
		48	9.25	- 320 + 40 x
		72	8.64	- 196.5 + 40 x
		96	7.75	- 260 + 40 x

Table 26.3: LC_{50} value of Dimethoate for 24 hr to *Anabas Testudineus* as calculated by *Logistic Response* model

X	*p*	*n*	*Y = loge P / (1-P)*	*W = np (1-P)*	*WX*	*WY*	*WX²*	*WXY*
.875	.20	10	- 1.38	1.6	1.4	- 2.208	1.225	- 1.932
.903	.40	10	- 0.405	2.4	2.16	- 0.972	1.950	- .877
.929	.60	10	0.405	2.4	2.22	0.972	2.062	.902
.954	.80	10	1.38	1.6	1.52	2.208	1.450	2.106
.977	*1.0*	*10*	*0*	*0*	*0*	*0*	*0*	*0*
				ΣW =8.0	ΣWX =7.3	ΣWY =0	ΣWX² =6.687	ΣWXY =1.99

x = Logarithm of concentrations

p = Proporation of Mortality

n = number of fish in each tank

$\overline{X} = \Sigma WX / \Sigma W = 7.3 / 8 = .9125$

$\overline{Y} = \Sigma WY / \Sigma W = 0 / 8 = 0$

$$b = \frac{(\Sigma WX)(\Sigma WXY) - (\Sigma WX)(\Sigma WY)}{(\Sigma W)(\Sigma WX^2) - (\Sigma WX^2)}$$

$= 0.031$ ....(i)

$a = \overline{Y} - b.\overline{X} = 0.0282$ .... (ii)

$(-a/b) = .909$

anti log .909 = 8.11

LC_{50} = 8.11 ppm

Table 26.4: LC_{50} value of Carbaryl for 24 hr to *Anabas Testudineus* as calculated by *Logistic Response* model

X	*p*	*n*	*Y = loge P / (1-P)*	*W = np (1-P)*	*WX*	*WY*	*WX²*	*WXY*
.977	.20	10	- 1.38	1.6	- 1.563	- 2.208	1.527	- 2.156
1.0	.40	10	- 0.405	2.4	2.4	- 0.972	2.4	- .972
1.02	.60	10	0.405	2.4	2.448	0.972	2.496	.991
1.04	.80	10	1.38	1.6	1.664	2.208	1.730	2.296
1.06	1.0	10	0	0	0	0	0	0
				ΣW =8.0	ΣWX =8.075	ΣWY =0	ΣWX² =8.153	ΣWXY =0.159

x = Logarithm of concentrations

p = Proporation of Mortality

n = number of fish in each tank

$$\overline{X} = \Sigma WX / \Sigma W = 8.075 / 8 = 1.009$$

$$\overline{Y} = \Sigma WY / \Sigma W = 0 / 8 = 0$$

$$b = \frac{(\Sigma WX)(\Sigma WXY) - (\Sigma WX)(\Sigma WY)}{(\Sigma W)(\Sigma WX^2) - (\Sigma WX^2)}$$

$$= 0.022 \qquad \text{........(i)}$$

$$a = \overline{Y} - b.\overline{X} = 0.0221 \qquad \text{.......(ii)}$$

$$(-a/b) = 1.0$$

anti log 1.0 = 10.0

LC_{50} = 10.0 ppm

Table 26.5: LC_{50} value of pesticidies to *Anabas Testudineus* calculated by Logistic Responses model

	Pesticides	*Time (hr)*	*Anti log*	*LC_{50} (ppm)*
1	Dimethoate	24	0.909	8.11
		48	0.880	7.58
		72	0.805	6.38
		96	0.709	5.11
2	Carbaryl	24	1.0	10.0
		48	0.958	9.07
		72	0.939	8.69
		96	0.882	7.62

Sublethal or Test Concentration

The formula of Hart *et al.*, (1945) when applied to the LC_{50} values of two pesticides, gave the sublethol dose of Dimethoate and Carbaryl as 2.02 ppm and 2.24 ppm respectively.

Behavioural Responses

Fish were exposed to various concentrations of the two pesticides to determine acute toxicity values for different intervals. The behavioural responses of the fish under the stress of individuals pesticide were recorded. The test fish exposed to different

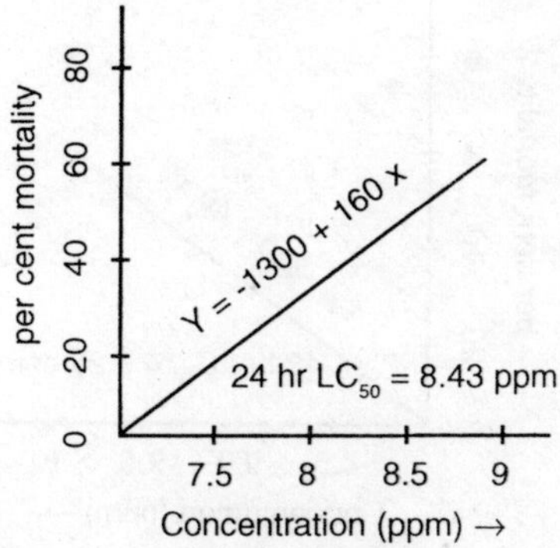

Fig. 26.1

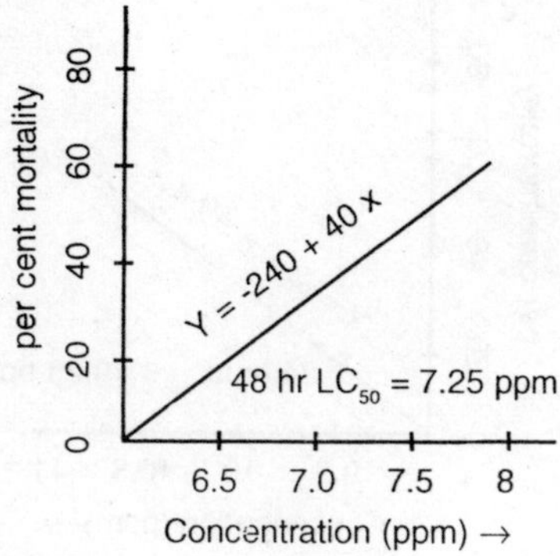

Fig. 26.2

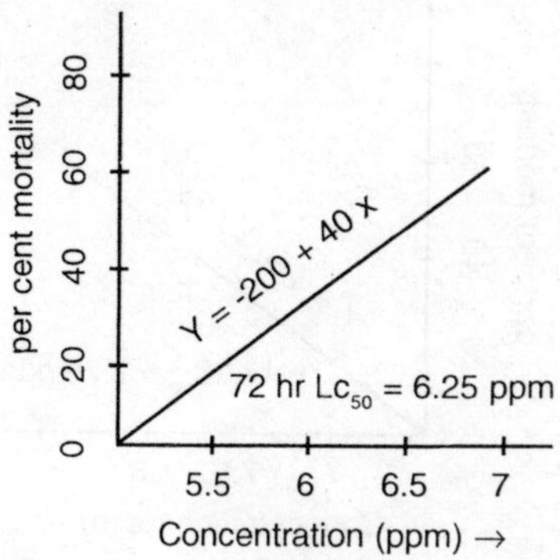

Fig. 26.3

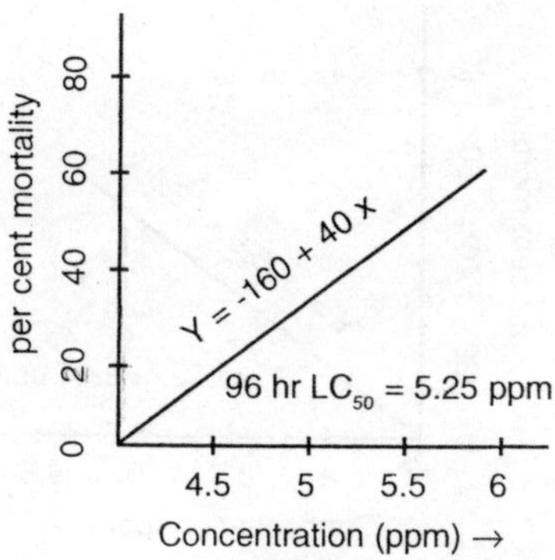

Fig. 26.4

Fig. 26.1 - 26.4 : LC_{50} values of carbaryl for 24, 48, 72 and 96 hours to the fish *Anabas testudineus* obtained by Regression Equation.

concentration of two pesticides exhibited more or less similar behavioural responses, which were more pronounced in case of Carbaryl exposure. The chief behavioural responses were faster and erratic opercular activity, surfacing and gulping of air, difficulty in respiration and convulsion. All these behavioural changes are reflection of signs of distress.

Discussion

It is evident that the toxicity of the toxicants was a functions of dosage and duration of exposure. The acutate toxicity of a toxicant is that at which rapid damage is caused to the organism by fast

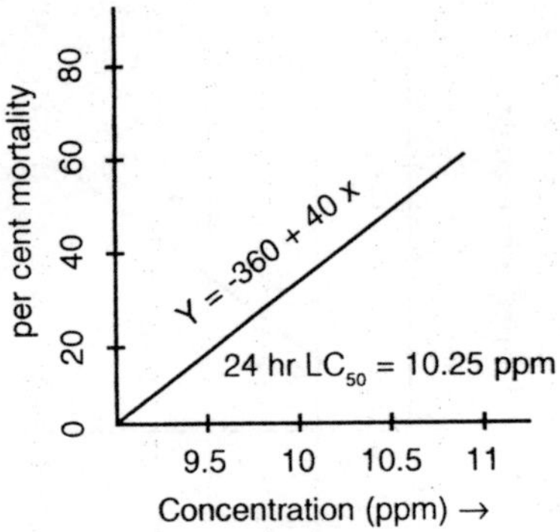

Fig. 26.5

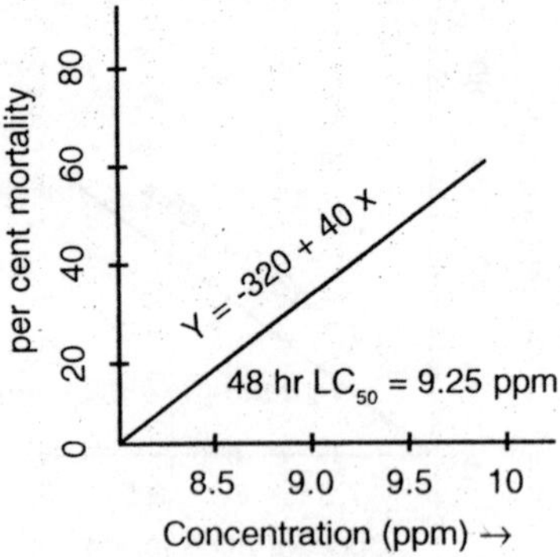

Fig. 26.6

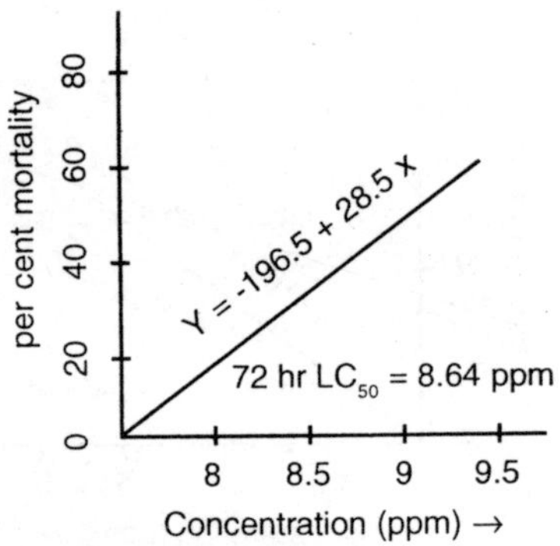

Fig. 26.7

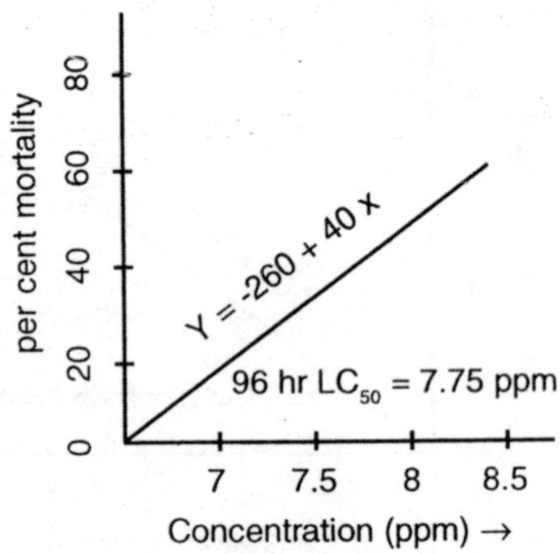

Fig. 26.8

Fig. 26.5 - 26.8 : LC_{50} values of carbaryl for 24, 48, 72 and 96 hours to the fish *Anabas testudineus* obtained by Regression Equation.

poisoning and it is fatal unless the organism escapes the toxic environment at any early stage (Sprague - 1969).

LC_{50} values obtained for different period of the two pesticides in the present study (Table 26.2 and 26.5). The relative toxicity of the two toxicant may be decided apparently, Carbaryl was found to be more toxic to Dimethoate in *Anabas testudineus*. The LC_{50} values of Dimethoate being 8.43, 7.25, 6.25 and 5.25 ppm at different time intervals (24, 48, 72 and 96 hr) as against 10.25, 9.25, 8.64 and 7.75 ppm respectively of carbaryl for the same period.

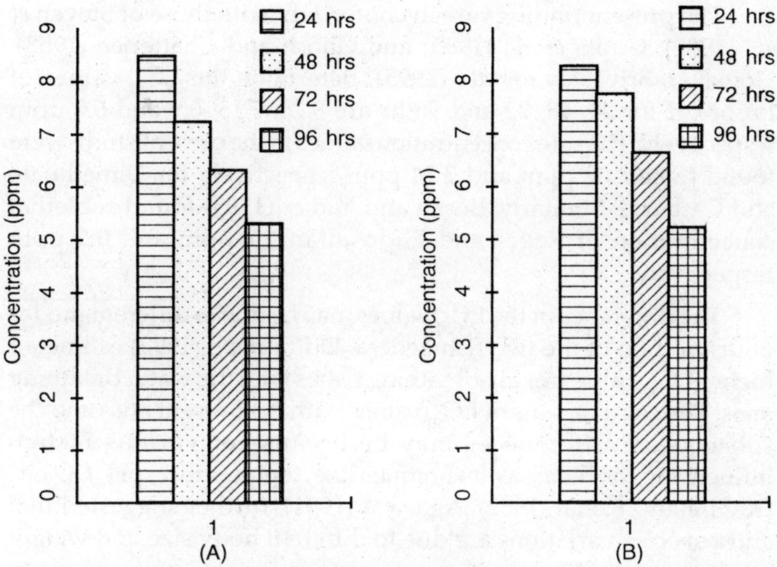

Fig 26.9: LC_{50} values (ppm) of Dimethoate for different time intervals obtained by Regression equation (fig.A) Logestic Response model (fig.B)

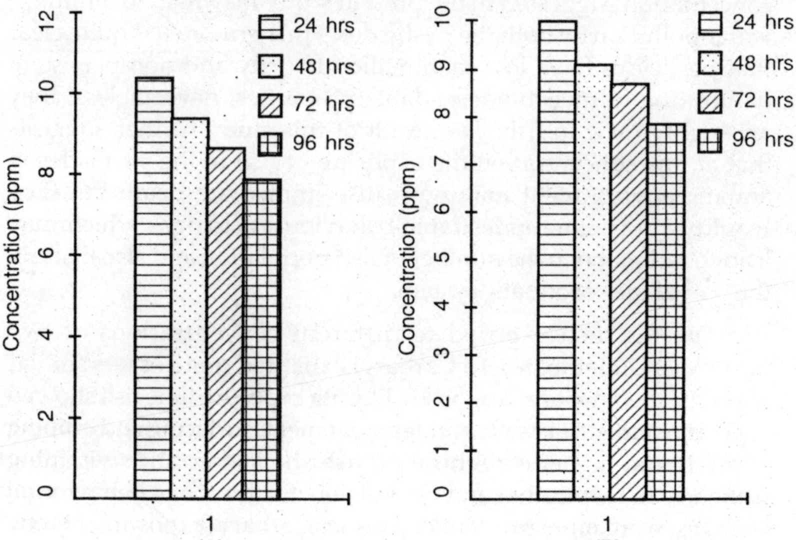

Fig 26.10: LC_{50} values (ppm) of Carbaryl for different time intervals obtained by Regression equation (fig.A) Logestic Response model (fig.B)

The present findings are in conformity with those of Steven *et al.*, (1985), Gupta *et al.*, (1985) and Ghosh and Chatterjee (1988). However, Srivastava *et al.*, (1995), determine the LC_{50} values of Propoxur for 24, 48, 72 and 96 hr are 8.25, 7.15, 6.9 and 6.5 ppm respectively. The safe concentration values in the present study were found to be 2.02 ppm and 2.24 ppm respectively for Dimethoate and Carbaryl. Similarly, Borah and Yadav (1995) found sublethal concentration of Roger and Endosulfan 1.7 ppm and 0.5 ppm respectively.

The difference in the LC_{50} values may be due to different modes of respiration of the two fish genera. Dutta *et al.*, (1992) estimated higher LC_{50} values for air-breathing fishes than the water breathing ones. The comparisons of LC_{50} values with the present ones and the subsequent differeneces may be because of various factors influencing the bioassay techniques like, temperature, pH, DO etc. (Kavira and Konar, 1982). Agarwal (1991), further suggested that entra species variations are due to different body size and weight as also the age of fish.

On the basis of present study it is clear that pesticides are hazardous to environment as well as aquatic lives. The safe concentation suggested in the present study may help to minimize water pollution by both the pesticides. Furthermore it is quite clear that pesticides have low mammalion toxicity and non-persistent nature due to their biodegradable properties, nevertheless, they are highly toxic to fish. The result of this study further suggests that at low concentration the exposure of fish to the pesticides in aquatic environment may not cause immediate death but they would bring some undesirable behavioural changes which may lead to reduction in the number of fish population and also disturb, the ecobalance of aquatic system.

The test fish exposed to different concentrations of two pesticides Dimethoate and Carbaryl exhibited more or less similar abnormal behavioural responses. During exposure time fish showed rapid movement, faster opercular movement, surfacing and gulping of air. But at higher concentration fish showed erratic swimming difficulty in respiration and convulsion. Such type of behavioural changes, were more pronounced under Carbaryl exposure thereby indicating greater toxicity of this chemical to the test fish. Due to profuse mucus secretion and congulation all over the body and gills, fish struggled for aerial breathing. This was followed by loss of equilibrium and fish slowly became lethargic and underwent hypoactivity. Similar behavioural responses have been reported by earlier workers under insecticides exposure and other such related pesticides (Ghosh, 1990; Bhatnagar and Tyagi, 1994 and Dutta *et al.*, 1994).

The large amount of mucus secretion all over the surface of the body and ultimately death occured as the results of exposure to pesticides. These findings are in conformity with the earlier findings of Pravaker *et al.*, (1993). Pandey *et al.*, (1990) further suggested that the heavy exudation of mucus over the body and body dispigmentation are attributed to the dysfunction of endocrine/ pitutary gland under the toxic stress. Sadhu (1993), suggested that opercular movement observed in the present study might be necessary to satisfy the requisite oxygen demand for survival. The observed mucus accumulation on the gills and skin of the fishes exposed to pesticides were probably due to toxic effects of pesticides became respiratory epithelium might be the main target site of toxicity during the period of experiment. Lat *et al.*, (1984) are also suggested that respiratory distress and increased opercular movements are the signs of oxygen depletion.

It is, therefore, conclude that the toxicity of pesticides depends upon a number of physical chemical and biological factors, each of which may be used as a tool for pesticide toxicity to fishes. In general temperature rises an increase in the toxicity, pH exerts its influence on ionnic compounds to determine the toxicity of a compound at any pH. The hardness on the toxicity of pesticides seems limited. Increased turbidity generally renders the pesticides less toxic.

References

APHA, AWWA, AWD, WPCF (1985). Standard methods for the examination of water and waste water. American Public Health Association, American Water Works Association and Water Pollution Control Federation. 16h Edition. American Public Health Association. Washington, D.C.

Agarwal, S.K. (1991). Bioassay evaluation of acute toxicity level of mercuric chloride to an airbreathing fish *Channa punctatus* (Bloch.): Mortality and behaviour study. J. Environ. Biol. 12(2): 99-106.

Annes, M.A. (1975). Acute toxicity of four organophosphorus insecticides to a freshwater teleost *Channa punctatus* (Bloch.). J. Zool. 7: 134-141.

Byron, W. and Brown, J.R. (1967). Quanto-response assay. In: Statistics in endocrinolcgy (Eds. J.W. Mc Arthur and T. Colton). MIT Press, Cambridge MA/London.

Borah and Yadav (1995). Static Bioassay and Toxicity of two pesticides. Rogar and Endosulfan to the air-breating fish, *Heteropneustes fossilis* with special references to behaviour poll. Res. 14(4): 435-438.

Bhatnagar, M.C. and Tyagi, M. (1994). Toxicity of Durmet and Mortal to a freshwater proc. Acad. Environ. Biol. 3(2): 177-180.

Choudhary, S. (1996). Studies on the adverse impacts of two newly formulated synthetic pyrithroids, Fenvalerate and Decamethrin, on various bio-chemical parameters of *Channa punctatus* (Bloch.). Ph.D. Thesis, L.N. Mithila University, Darbhanga.

Dowrie, N.M. and Heath, R.W. (1970). Basic statistical methods, Harper international New York, Evanston and London, 129-131.

Dutta, H.M., Nassar, S.S.T., Munshi, J.S.D. and Richmonds, R.C. (1994). Behavioural charges in an air-breathing fish, *Anabastestudineus*, exposed to Malathion Bull. Enrich. Contan. Toxicol. 52, 80-86.

Gupta *et al.*, Gupta, H.C., Singh, R., Jain, K.L. and Saxena, R.C. (1985). Relative and residual toxicity of some synthetic pyrethroids *Pestology* 9: 39-42.

Ghosh, T.K. and Chatterjee, S.K. (1988). Toxic influence of Fenvalerate on to Biological Parameters of the fish *Anabas testudinas*. Enrich, Ecol. 6(1): 107-110.

Ghosh, T.K. (1990). Toxicological evaluation of synthetic pyrethroids on the freshwater fishes. In: Environmental pollution and health hazourds (eds. R. Prakash and S.M. Choubey). Society of Bio Sciences, Muzaffarpur, India. pp. 15-25.

Hart, W.B., Doudoroff, P. and Greenbank, J. (1945). The evaluation of toxicity of industrial wastes, chemicals and other substances to fresh water fishes. Atlantic Refinii,g co. Pill. Part (I): 317-326.

Jacob, S.S., Balakrishnan Nair, W. and Balasubramanian, N.K. (1982). Toxicity of certain mosquito larvicides to the larvivorous fishes Aplo cheilus lineatus (Cur and Vel) and Macropodus supanus (Cuv and Vel) Enrich. Pollut. 28, 7.

Johnson, W.W. and Finlay, M.T. (1980). Hand book of Acute toxicity of chemicals to fish and aquatic invertebrates. Resouce Publ. 137. Fish and wild life service. U.S. Department of the interior Washington, D.C.

Kaviraj, A. and Konar, S.K. (1982). Acute toxicity of Mercury, Chromium and Cadrius to fish, Plankton and Warm Geobios 9(3): 97-100.

Konar, D.S. (1981). Pollution of water by pesticides and its influence on aquatic ecosystems. Ind. Rev. Life Sc. I, 139-165.

Mehrotra, K.N. (1990). Pest control strategies for 2000 A.D. *Proc. Ind. Natn. Sci. Acad.* B52(1): 10-16.

Pandey, A., Kunwar, G.K. and Mushi, J.S.D. (1990). Integumentary chromatophores and mucus glance of fish as indicator of heavy metal pollution. J. Freshwater Biol. 117-121.

Pravakar, J.D., Martin, R.E., Jagpat, N.R. and Kshemikalyani, S.B. (1993). Effect of indosulfan and Hexachlorocyclohexane (HCH) on the behaviour of fish *Labeo rohita*. J. Eco. Biol. 5(2): 149-150.

Sadhu, D.N. (1993). Toxicity of an organophosphorous insecticide monocil to the air-breathing fish, *Channa punctatus*, J. Eco. toxicol. Environ. Mont. 3(2): 133-136.

Shafi, S.M. and Alam, M.N. (1990). Pollutional potential of an insecticide mattacid in respect of pond water fish food organisms. Nat. Sem. Fish and their environment Sept. 19-22, Unive. of Kashmir.

Srivastava, A.K., Singh, N.N. and Srivastava, A.K. (1995). Biochemical changes in freshwater Indian catfish (*heteropneustes fossilis*) following exposure to sublethal concentration of propoxur. J. Fresh water Biol. 7(4): 252-260.

Sprague, J.B. (1969). Measurement of pollutant toxicity to fish I. Bioessay methods for acute toxicity. Review paper water Res. 3: 739.

Swarup, P.A., Rao, D.M. and Murty, A.S. (1981). Toxicity of endosulfan to the freshwater fish, cirrhinus mrigala. Bull. Enviorn. Contam. Toxicol. 27: 850-855.

Steven, B., Joet, P. Coasts, R. and Mckin, J.M. (1985). Differential toxicity and uptake of 2 fenvalerate formulations in fathead minnows. *Pinephles, promelas, Environ. Toxicol. Chem.* 4: 533-542.

27

Ionic Imbalance as a Consequence of Chromium Exposure in the Catfish *Heteropneustes Fossilis* (Bloch)

D.N. Roy
Lecturer in Zoology,
M.L.S.M. College,
Dharbhanga, Bihar

N.K. Dubey
University Department of Zoology,
L.N. Mithila University,
Dharbhanga, Bihar

Abstract

The ion balance in the catfish (*Heteropneustes fossilis*) was studied after 96 hours, 15 days, 30 days and 45 days of exposure to sublethal concentration of trivalent chromium (9.945 mg/g/l) in fresh water. The trivalent chromium (Cr^{+3}) had no effect on plasma potassium level but plasma sodium and plasma calcium showed depletion in 96 hour of chromium exposure. In 15 days of exposure, only sodium showed depletion but levels of plasma potassium and plasma calcium were found elevated. In 30 days and 45 days of Cr^{+3} exposure, levels of sodium potassium and calcium of plasma showed a definite pattern of elevation.

Hyponatraemia followed by dilutional hypermatraemia; normokalaemia followed by hperkalaemia and hypocalcaemia followed by hpercalcaemia point towards disturbed osmoregulation and impairment of ion-regulating tissues - like kidneys and gills.

Introduction

The more or less complete constancy of the internal environment is an extremely important factor which enables fishes to adapt themselves to living in waters of various salinities. This constancy is attained by development of various osmoregulatory adaptations which enables the fish to regulate the osmotic pressure of its body fluids. The more perfect the osmoregulatory adaptations, the greater the differences between the composition and pressures of the internal fluids of the organism and its external environment.

The effects of dissolved compounds which enter the water both organic and inorganic are variable. Trivalent chromium (Cr^{+3}) for example, is a common contaminant of surface waters (Pickering, 1980; Sloof *et al.*, 1989). Concentration of chromium in the micro molecular range have been reported frequently in rivers and lakes (Beg *et al.*, 1990; Aveset *et al.*, 1993; Zang *et al.*, 1994; Samanta, 1999).

The metal may either be present in trivalent form (Cr^{+3}) or in hexavalent form (Cr^{+6}). The later is considered the most toxic form of aquatic organisms (Vander putte *et al.*, 1981). Water pH is one of the most important characteristic determining chromium toxicity to fish (Vander putte, 1980). The dominant species of Cr^{+3} are $Cr(OH)^{++}$ at pH 4; $Cr(OH)_2^{+}$ at pH 6 and $Cr(OH)_3^{2}$ at pH 8 (Stern, 1982). Cr^{+3} is largely associated with particulate matter subject to sedimentation and filtration making it less bio-available.

The main function of electrolytes are maintenance of proper-osmolarity and volume of body fluids, maintenance of acid-base equilibrium, exert special physiological effects of certain ions ex. g-effect of Ca^{+2} ion in neuro-muscular excitability.

Present study of plasma ion-balance was carried out as part of the experiment to explore possible impact of plasma-ions (Na, K, Ca) in HCT, Hb per cent plasma glucose and total plasma protein levels in Cr^{+3} exposed fish *H. Fossilis.*

Material and Methods

Fish were procured from local ponds and market and transported to laboratory for acclimation. They were fed and maintained as per suggestion of APHA, AWWA, WPCF (16th ed., 1985).

The test chemical used in this experiment was chromium sulphate (Burgyne Chemical Ltd., Calcutta). $[Cr_2(SO_4)_3.4H_2O]$. Test concentrations were prepared in tap water (Average temperature $23 \pm 3^{\circ}C$; pH 7.6 ± 0.2; dissolved oxygen 6.8 ± 1.2 mg/l; free CO_2 1.3 $\pm$ 1.7 mg/l; total hardness as $CaCo_3$ 185 ± 1.5 mg/l and total alkalinity as $CaCo_3$ 135 ± 3.5 mg/l.

Feeding was however stopped 24 hours before the start of experiment. A static bio-assay test was performed to determine the 24 hr, 48 hr, 72 hr and 96 hr LC_{50} of Chromium sulphate (Cr^{+3}) to *H. Fossilis* following the methods of APHA, AWWA and WPCF (16th ed. 1985). The sublethal concentration was obtained by employing the formula of Hart *et al.*, (1945) using 0.3 as application feator.

Four aquaria (A,B,C,D) each filled with 10 L tap water and 10 acclimated fish (L = 12-14 cm, W = 16-18 g) accompanied with their respective control were exposed to sub-lethal concentration of Cr^{+3} (9.945 mg/g/l).

Exposure medium was renewed every 24 hours to maintain the effective concentration. However, treated fish were allowed to feed on equal amount of finely chopped viscera of chicken obtained from local poultry shop in long-term exposure.

At the end of exposure period (96 hr, day 15, day 30 and day 45) the treated and control fish were anaesthetized by ms 222 (Tricane, methane, sulfonate, sandoz) for two minutes. The blood samples were collected in heparinized plastic vials with the help of 1 ml disposable syringe equipped with 22-gauge hypodermic needle by caudal peduncle cut. Blood obtained from individual fish was centrifuged (3000 rpm for 5 mts) and plasma was collected in different vials and stored at 20°C until analysed.

Plasma sodium, potassium and calcium were measured by Flame emission photometry after (Wotten, 1974) using flame photometer.

Results and Discussion

Results are shown in Table 27.1

In fish, chromium is considered as a substance of average acute toxicity (Peres and Brichon, 1980) Chromium could enter the fish by diffusion across the gill membrane as postulated by Knoll and Fromm (1960). The dietary absorption of Cr^{+3} in vertebrates is less than 1 per cent regardless of the dosage level. The acid pH in the stomach keeps the chromium in Cr^{+3} (Trivalent form) state and also reduces Cr^{+6} to Cr^{+3}, but once in small intestine, the alkaline pH will convert the Cr^{+3} to Cr^{+6}. Since the absorptivity of Cr^{+6} is very much greater than the trivalent form, Cr^{+6} is estimated to be 100 times more toxic than Cr^{+3} (Wong. M.H. 1982).

Table 27.1: Plasma ion (Na, K, Ca) concentration in control and Cr^{+3} exposed fish *H. Fossilis*

Electrolytes	*Control*	*Treated*			
		96 hr	*15 day*	*30 day*	*45 day*
Na^{+} µM/L	125.1	124.22 (< 0.7%)	121.98 (< 2.49%)	132.48 (> 5.9%)	135.35 (> 8.2%)
K^{+} µM/L	2.11	2.11	2.12 (> 0.7%)	2.12 (> 0.8%)	2.19 (> 3.79%)
Ca µM/L	4.0	3.4 (< 15%)	4.04 (> 10%)	4.5 (> 12.5%)	4.34 (> 8.5%)
Total (Na+K+Ca)	*131.12*	*129.73 (< 1.12%)*	*128.14 (< 2.33%)*	*139.1 (> 6.0%)*	*142.04 (> 8.25%)*

n = 3 in each case expressed as µM/L

It is believed that Cr^{+3} enters the circulatory system of *H. Fossilis* during respiration as is then carried to the liver. Together with the bile, the chromium is released into the gut and causes severe impairment to the intestine. However, it also damages the gill tissues causing formation of a film of coagulated mucus, bringing about death by asphyxiation; the changes is similar to the effects of Zn and Cu salts on carps and several other fishes.

In the present study 1.12 per cent total electrolyte depletion in 96 hr exposure have been observed with individual contribution of sodium about 0.7 per cent and calcium about 15 per cent delepted. Potassium, however, remained unchanged. But in 15 days of exposure percentage of ion distribution and their concentration is peculiar. Total electrolyte depletion is 2.33 per cent with the depletion of 2.49 per cent sodium depletion counter balanced by 0.7 per cent potassium and about 10 per cent calcium elevation.

Thus, 96 hr. exposure suggest haemoconcentration due to urinary loss of water and 15 days of exposure suggest tissue dehydration with increased level of plasma salts to retain water under the influence of salt retaining hormone mineralo-corticoid. Slightly elevated level of potassium (0.7 per cent) in 15 days is suggestive of cell and tissue damage. In 96 hr. exposure depletion of calcium ion is suggestive of neuromuscular excitability and increase in 15 days of exposure is suggestive of increased absorption of calcium by gills from exposure medium.

Changes in the functioning of gills occurs when the fish are stressed by hypoxia (due to gill damage) which might be expected to increase the permeability of the gills to water influx (Randall *et al.*, 1972). Also, the gills are the primary site of extra-renal-ion regulation (conte, F., P. 1969) and the ion/water balance, there will be disturbed and could increase in urine flow and a consequent change in the renal-ion-excreation pattern.

According to Swift and Lloyd (1974) elevated urine flow rates and concomitant increased haematocrit followed by lower than normal urine flow rates suggest that haemoconcentration is the initial response to anoxia. If the change in water balance is a reflection of change in water volume, then there is an initial haemoconcentration followed by a tendency of returning blood volume to its normal level. The simplest relationship between HCT and urine filtration rate (UFR) would be an initial diuresis with haemoconcentration giving an increased HCT value.

Chromium has been shown to have a strong relationship with plasma sodium. Levels of Na^+ in ECF and intracellular fluid is generally maintained along with glucose by a mechanism of uniport and co-transport (symport and uniport) as devised by Albert B. *et al* (1983) (Quoted in Harper's Biochemistry). Osmotic function of glucose causes variation in the levels of plasma sodium. And, it is worth mentioning here that hyperglycemic response is the commonest menifestation of Crtoxicity in fish.

Factors such as nutrition and endocrine influences, sexual differences and stress are possibly of importance imparting ionic balance of the blood plasma (Fange and Lidman, 1976). The tendency towards slightly elevated sodium level (dilutional hyper natraemia) may be due to an active renal absorption of sodium in combination with leak of external excretion of sodium (Mc. Inerney, 1974).

But Kunhert *et al.*, (1976) opined that Na^+/K^+ at phase activity in intestine and renal tubules becomes impaired by chromium. It is, therefore, expected that the mucus layer on the gill surface creates a micro-environment which may act as an 'ion-trap'. This might delay the penetration of the heavy metal into the underlying living system (Part, 1983).

Zacon *et al.*, (1989) suggested that the greatest significance of increasing mucification might be protection from the irritating agents present in external milieu. They also concluded that the greatest abundance of mucus cells might be related to the maintenance of ionic equilibrium since some glycoproteins may function in attracting cations (+ve ion) from the environment and in retarding cation efflux from fish during condition of stress. Cr^{-}induced lowering of sodium have also been reported by Vander putt *et al.*, (1981).

Except in 96 hour of exposure, levels of potassium were found elevated in 15 days, 30 days and 45 days of Cr^{+3} exposure in *H. Fossilis*. Larson and Haux (1981) have also reported imbalance of potassium and calcium in cadmium exposed flounder. Soivio and Oikari (1976) observed erratic changes of potassium in Esox lucius under stress. In the present investigation Cr^{+3} induced potassium elevation in short-term exposure might be related to stress. Its persistence in the long-term exposure might be due to high range of variation. This may in part be due to a leakage of potassium ions from cells affected by stress, low oxygen or otherwise disturbed (Fange and Lidman, 1976). In all vertebrates levels of potassium ion remains low indicating its excretion by kidney. Since potassium remains mostly within the cell, the break down of cells causes its rise in blood and urine. Therefore, monitoring of potassium ion serves as a good guide as to the degree of disintegrating cells of the body.

A rise of plasma potassium not only indicates tissue damage but also lowering of pH *i.e.* acidosis. For every 0.1 pH unit decrease serum potassium will rise approximately 0.6 m Eq/L and conversely (Schrierr R.W. 1976). The kidney is the main channel of H+ ion execreation and its failure leads to acidosis.

The depleted level of calcium in 96 hr Cr^{+3} exposure shows increasing trend in 15 day, 30 day and 45 days of exposure. In general, fresh water fish do not require a large amount of calcium or phosphate for use in skeletal gravitational support. In addition to low skeletal requirement for calcium ion, the fresh water fishes are confronted with a lower environment than marine fish although net Ca^{+2} uptake is higher than in sea water.

In short-term exposure perhaps due to neuromuscular disturbances levels of calcium decreased but in long-term exposure more calcium ion detected in plasma, might be due to more calcium ion uptake by gills, as the first cause of elevation of plasma calcium level in Cr^{+3} exposed fish *H. fossilis*. Most probably, in long-term exposure failure of damaged intestine by chromium to absorb more calcium causes vitamin D resistance in fish responsible for raised levels of plasma calcium in Cr^{+3} exposed fish *H. fossilis.*

The physiological control of calcium in fish is little known. True fishes are provided with calcitonin producing ultimobranchial bodies, but parathyroid glands have not been detected (Copp, 1969). The livers of many fishes contains enormous amount of vitamin D, but it is not known if this substance has any hormonal calcium regulating function as in higher vertebrates.

Schrier, R.W. (1976) reported that in decreased renal function one finds hypocalcaemia with subsequent development of hypercalcaenia. Decreased renal function is associated with more phosphate retention with vitamin-D resistance with resultant decreased intestinal absorption of calcium and hypocalcaemia. Vitamin-D resistance is due to the fact that the kidney normally convert 25-hydroxy cholecalciferol to 1, 25-$(OH)_2$ chole calciferol which is the most potent metalbolite of vitamin-D (De Luca H.F. 1974).

After intestinal damage and tubular damage by Cr^{+3} it is the muscle which appears to be the most suitable target by Cr^{+3} toxicity and in such a situation the muscle suffers much by liberating more calcium ion to provide active sites for the binding of heavy metal chromium, with known affinity for proteins and ligands to bind with. Released calcium ions finds space in circulating fluid.

Conclusion

Trivalent chromium, thus, appears to cause, profound ion-imbalance in the catfish *H. Fossilis* both in short-term and long-term exposure. These alterations in plasma ion concentrations point towards imparied function and alterations in ion-regulating and osmo regulating tissues of the experimental fish.

References

APHA, AWWA and WPCF (1985). Standard methods for examination of water and waste water. American Public Health Association, American Water Works Association and Water Pollution Control Federation. 16th edition, American Public Health Association, Washington D.C.

Anti, Soivio and Aimo, Oikari (1976). Hematological Effects of Stress on Teleost, ESOX luciusl. J. Fish. Biol. (1976): 8, 397-411.

Alves, M.M., C.G. Gonzalez Beca, R.G. Guedes de Car valho, J.M. Castanheiva, M.C.S. Pareiva and L.A.T. Vasconcelos (1993). Chromium Removal in Tannery Waste Waters "Polishing" by Pinus Sylvestris bark. Water. Res. 27(8): 1333-1338.

A.K.E. Larson, Bengt. Evik Bengtsson and Carl Haux. (1981). Distrubed Ion Balance in Flounder, Platichthys Flesus L. Exposed to Sublethal levels of Cadmium. Aquatic Toxicology, 1(1981): 19-35.

Beg, M.A.A., S.N. Mahmood and S. Naeem (1990). Pollution Due to Tannery Effluents in the Korangi Industrial Area, Karachi (Pakistan), Pakistan J. Sci. Indust. Res. 33(10): 431-435.

Copp, D.H. (1969). The ultimo Branchial Glands and Calcium Regulation. In Fish Physiology II (Eds. W.S. Hoar and D.J. Randall) pp. 377-394, New York and London: Academic Press.

Conte, F.P. (1969). Salt Secretion in Fish Physiololgy (Eds. Hoar, W.S. and Randall, D.J.) Vo. 1, pp. 241-292, New York, Academic Press.

Deluca, H.F. (1974). Vitamin D. Amer. J. Med. 57;1, 1974.

Fromm, P.O. and R.H. Schiffman (1958). Toxic Action of Hexavalent Chromium on large mouthbass. J. Widll. Manage. 22, 40-44.

Hart, W.B., Doudoroff, P. and Green Bank, J. (1945). The evaluation of toxicity of industrial wastes, Chemical and other Substances to fresh water fishes. Atlantic Refining Company, Philadelphia, 317-326.

Kunhert, P.M., B.R. Kunhert and R.M. Stokes (1976). The effects of invivo chromium exposure on Na-K and Mg ATP as activitiy in several tissues of rainbow trout (Salmo gairdneri) Bull. Environ. contm. Toxicol. 15: 383-390.

Knoll, J. and P.O. Fromm. (1960). Accumulation and Elimination of Hexavalent chromium in Rainbow Trout.Physiol. Zool. 33: 1-8.

M.H. Wong, W.M. Lau, T.Y. Tong, W.K. Liu and K.L. Luk (1982). Toxic Effects of Chronic Sulphate on the Common Camp Cyprinus Carpio. Toxicology Letters, 10(1982): 225-232.

M.C. Inerney, J. E (1974). Renal Sodium Reabsorption in the Hag fish, Eptatretus Stouti Comp. Bio. Chem. Physiol. 49: 273-280.

Part, P. (1983). Cadmium Uptake in Perfused Rainbow Trout Gills. Acta Universitatis Upsaliensis, Dissertation from Faculty of Science. 707.

Part, P. and Lock, R.A.C. (1983). Diffusion of Calcium, Cadmium and Mercury in a Mucous Solution from Rainbow Trout. Comp. Biochem. Physiol. 76: 259-263.

Pickeying, Q.H. (1980). Chronic Toxicity of Hexavalent Chromium to the Fathead Minnoiw Pimephalus Promeales. Arch. Environ. Cotaim. Toxicol. 9: 405-413.

Peres, G. and M.H. Brichon (1980). Etude Comparative au Moyend untest Statique (AFNOR Sur 24 h) deal Sensibilite Relative de plusieurs especes de poissons au bichromate depotassium. Rapport du Symposium International de Sophia-Antipolis, pp. 320-328.

Randall, D.J., Baumagarten, D. and Malyusz, M. (1972). The relationship between Gas and Ion Transfer across the Gills of Fishes. Comp. Biochem. Physiol. 41: 629-637.

Ragnar Frange and Ulf Lidman (1976). Comparative Studies of Inorganic Substances in the Blood of Fishes from the Scagerace Sea. J. Fish. Biol. (1976): 8, 441-448.

Sloof, W., R.F.M.J. Cleven, J.A. Janes and P. Vander Poel (1989). Basic Document Chromium. RIVM. report no. 758701001 Bilthoven, 146 pp.

S. Samanta (1999). Heavy Metals as an indicators of Stress Condition of the Ecosystem. Short Course of Training on Aquatic Environment Impact Assessment. CICFRI (ICAR) Bull. No. 88. Jan. 1999.

Sterh, R.M. (1982). Chromium Compounds: Production and Occupational Exposure in Biological and Enviornmental Aspects of Chromium, edited by S. Langard. Elsevier Biomedical Press. Amsterdam. Chapter - 2 pp. 5-48.

Schrier, R.W. (1976). Renal and Electrolyte Disorders Little Brown and Co. Boston 1976.

Vander putte, I., J. Lubbers and Z. Kolan (1981). Effect of pH on Uptake, Tissue distribution and Retention on Hexavalent Chromium in Rainbow Trout (Salmogairdneri) Aquat. Toxicol. I: 3-18.

Wedmeyev, G. and W.T. Yasutake (1977). Clinical Methods for Assessing the Effect of Environmental Stress on Fish Health. U.S. Fish Wildl. Serv. Tech. Paper. 89, 18 pp.

Zacone, G., Lo Cascio P., Fasulo, S. and Licata A. (1985). The Effects of an anionic Detergent on Complex Carbohydrates and Enzyme Activities in the Epidemis of the Catfish Heteropneustes fossilis (Bl.). Histochem. J. 17: 453-466.

Zacone, G., Fasulo, S., Lo Cascio P. Ainis, L. Ricca, M.B. and Licata A. (1989). Effects of Chronic Exposure to endosulfam on Complex Carbohydrates and Enzyme Activities in Gill and Epidermal Tissues of the Fresh Water Catfish *H. fossilis* (Bl.). Arch. Biol. (Bruxelles), 100: 171-185.

Zhang, J., W.W. Huang, M.G. Liu and J.Z. Cui (1994). Eco-Social Impact and Chemcial Regimes of Large Chinese Rivers: A Short Discussion. Water Res. 28(3): 609-617.

28

Anaerobic Treatment of Organic Wastes

D.R. Khanna and Praveen Sarkar
Department of Zoology and
Environmental Sciences,
Gurukul Kangri University,
Hardwar (UP)

Ashutosh Gautam
India Glycols Ltd.
A-1, Industrial Area,
Bazpur Road,
Kashipur - 244 713 (UP)

Abstract

In the recent years there has been a growing concern over the problem of municipal and industrial waste disposal. In India, city sewage, distilleries, tanneries, breweries, sugar mill and food processing units are the major source of organic pollution in water resources. Our growing concern about the availability of clean environment/water makes us more concious to develop effective and beneficial treatment technologies for the liquid wastes. Anaerobic treatment of liquid waste, having high concenration of biodegradable organic matter, is a competent method to overcome this problem. With the aid of anaerobic treatment technology such wastes can be used to extract useful products/byproducts and energy, like-biogas. Biogas is a clean and effective substitute of other conventional energy sources, *i.e.* coal. The article highlights major issues of anaerobic treatment alongwith the cost economics.

Introduction

Water is the elixir of life. In our day-to-day life we are using this vital natural resource for a variety of purposes, like drinking and other domestic uses, irrigation, power generation, industrial cooling and finally for waste disposal. Continuous practice of our society to use water resources as a natural dustbin is creating the greatest problem of this century - The water pollution. Water pollution may be of different types depending upon the constitutents causing pollution, for example, organic pollution is mainly caused by the discharge of organic wastes or wastes containing organic material. Such wastes when enter a water body, dissolved oxygen is consumed in their oxidation or breakdown by micro-organisms. The more the oxygen required for oxidation, the greater will be the deoxigenation, which can cause adverse impacts on the existing flora and fauna of the water body. Pollutants in the category of organic pollution generally comes from such sources as : domestic/city sewage, food processing units, distilleries, pulp and paper mill, sugar mill, breweries, dairy, slaughter house, etc. The water pollution problems in developing countries, like India, are still at a stage where effects can be abate if action is taken in time. Always, it is easier and advantageous to control pollution at the early stages rather than investing huge capital to protect and purify the water resources at later stages like - Ganga Action Plan.

Waste

Waste is the disposal/discharge of material, residues and energy into the enviornment. Of these, some are unconvertible raw materials, some are unrecovered final products and some are by products. Thus discharge of waste in the environment creates a problem of environmental pollution. In other words it can be considered that we are loosing by two ways. Firstly, we are loosing our valuable material and secondly we are polluting our environment, thereby our natural resources.

Liquid waste containing biodegradable organic matter can be used to produce some useful products/energy like - methane rich fuel gas commonly termed as BIOGAS.

Water Pollution Control

On the discharge of untreated organic waste to aquatic environment —

(i) Dissolved oxygen level of receiving aquatic body diminishes and results into subsequent damage to the aquatic organisms.

(ii) High turbidity hinders the penetration of sunlight which adversely affects the photosynthetic activity.

(iii) Floating impurities and foul smell due to the presence of organic matter makes water unsuitable for recreational purposes.

(iv) Water no longer remain suitable for domestic/irrigation purposes.

(v) Bioaccumulation/Biomagnification/Bioamplification of toxic material takes place which ultimately attack the human population.

Therefore, the control of water pollution is essential and require instant attention and devotion. The most effective way to control water pollution is the treatment of waste water in an efficient and suitable effluent treatment plant before its discharge. The main objective is to minimize harmful constituents to an acceptable level with or without by-product recovery. Various suggested methods are —

(i) Physico-chemical methods,

(ii) Biological treatment methods,

(iii) Incineration, and

(iv) Production of Green Manure (composting).

Physico-chemical methods are generally used in separating the heavy materials and to neutralise the acidity of basicity. These are - Screening, Sedimentation, Floculation, Neutralization etc.

The objective of biological treatment is to remove oxygen consuming organic matter by the help of micro-organisms, capabale in oxidising the organic matter. Biological treatment is often the most competent methods.

These are of two types:

Aerobic: micro-organisms require oxygen to oxidise the organic matter and produce carbon dioxide as a by-product.

Anaerobic: micro-organisms does not require oxygen and produce biogas as a by-product.

Incineration is the drying of organic wastes and the dried material may disposed off on land.

Liquid organic waste can be utilised to produce green manure by treating with solid wastes (composting). The production of green manure with this method is highly beneficial and the manure produced is always better than the chemical manure.

Anaerobic treatment of organic wastes

Anaerobic digestion with methane recovery is indeed the most rational approach in present times of energy crisis. This treatment of organic wastes is also known as methanization or methane fermentation. Methane generation from the rotting of organic matter has been known since the 18th century. Volta identified this as marsh gas. The first use of this valuable gas as a fuel was recorded in 1875, when it was used for lightening the street light of a city in England. Due to various advantages (with few drawbacks) this method has gained a paramount interest and widespread acceptability.

There are several technologies available in the market to produce biogas from organic wastes. Process design alternatives for anaerobic treatment of organic wastes may be identified as anaerobic filter process, anaerobic single stage process and acid methane segregation process.

Anaerobic filter process has been developed to treat strong wastes, having high concentration of organic matter. The reactor contains a solid support, usually plastic matrix or stones. In this process micro-organisms grow attached to the support material or as free blocks in the interstices.

Single stage anaerobic process is a conventional anaerobic treatment process. This a combined one step process of all activities (acid formation and methane formation) required to recover methane rich fuel gas from organic waste. In this process only a single reactor is required.

Acid methane segregation process is a microbial segregation process where acid producing and methane producing micro-organisms are allowed to grow in different reactors under different environmental conditions. This process enhanced the activity of micro-organisms to utilise organic compounds more effectively.

Identification of organic wastes for anaerobic treatement

The liquid waste have high concentration of biodegradable organic material with adequate nutrients and micro-nutrients are quite suitable for anaerobic treatment and can be utilized to produce biogas. These wastes should be free from toxic substances. In terms of BOD the wastes from distillery, starch, pharmaceutical, city, pulp and paper, slaughter house, tannery, brewery, sugar mill etc. are highly suitable for the production of biogas with the help of anaerobic treatment. It is essential to know the status of organic waste before adopting the technology.

Process parameters for anaerobic treatment

Every process depends upon several environmental conditions. For the anaerobic treatment termpature, pH, nutrients and the presence of toxic substance are important.

Temperature

The most suitable temperature range is 30-40°C. In the lower range of the termpature the process remain slow due to low activity of the micro-organisms, whereas in the higher range it initially touches the peak but after that it slows down due to the sudden fall in the acitivity and the number of micro-organisms. Sudden variations in temperature results in temporary shut of the activity, but a long hold can cause degradation in the treatment process. On the basis of temperature range the process and the microorganisms are of three types *i.e.* cryophylic (between 0-25°C), conversion rates are slow. Mesophilic (between 30-40°C) and Thermophilic (between 50-70°C), the conversion rates are high due to high microbial activity. Normally the mesophilic range of temperature is most suitable and cost effective for the anaerobic treatment.

pH

The most suitable pH range for anaerobic treatment is 7.0-7.5, but is can be operated upto 8.0, the experimentation shows that in the higher pH range biogas have increased concentration of methane. Of the two bacteria, involved in anaerobic treatement, acidogens are capable of performing their activity between 5.0-6.0 pH range whereas 7.0-8.0 is suitable for methanogens.

Nutrients

Various studies showed that organic waste must be rich in nutrients and micronutrients for better anaerobic treatment. The main nutrients are N and P. In different technologies the rate of nutrients varies as per the COD/BOD loading rate.

Toxic substances

Like other biological process anaerobic treatement may also be affected by the presence of toxic substances. For this process sulphide, ammonia and higher fatty acids are the most consquential. These compounds always try to inhibit the microorganisms' activity which ultimately end in the lowering of BOD/COD reduction and biogas production.

Advantages of Anaerobic Treatment

Anaerobic treatment process has several advantages from other biological processes. Few important are discussed hereunder. This can accept high organic loading since oxygen requirement is nil.

(i) This produces a useful by-product *viz.* Biogas.

(ii) This can be operated after along non-operating gap.

(iii) Since the reactor is closed no offensive odour exists.

(iv) Sludge production is low.

(v) Complete destruction of pathogens.

(vi) Possible direct utilization of sludge as a manure.

(vii) Treatment effluent can be used for aquaculture.

The main advantage is the production of biogas, which can be used a clean and enviornment friendly fuel.

Biogas

Biogas is a composite mixture of methane, carbon dioxide and hydrogen sulphide produced naturally by the decomposition of organic matter. The composition of all the three gases differ from time to time and technology to technology. The main composition is – Methane (CH_4) - 50-70 per cent, Hydrogen Sulphide (H_2S) - 2.5 per cent and Carbon Dioxide (CO_2) - rest. The calorific value of the biogas is approximately 5000 K.cal/NM^3. Biogas is a good substitute of conventional energy sources like wood, coal, kerosene etc. Biogas production from organic wastes varies from waste to waste. Generally, it ranges from 0.1 M^3 to 0.55 M^3/Kg. of COD reduced.

Benefits from Biogas

There is an assumption to know exactly how much biogas is benefitial to us. Let us consider a strong waste *i.e.* waste from distillery

(Spentwash). 1 M^3 of distillery waste can produce 25-35 NM^3 of biogas. (The variation is due to technology difference).

Methane percentage	-	70 per cent
Calorific Value	-	5000 K.cal/NM^3
Calorific value of coal	-	4500 K.cal/NM^3
Combustion efficiency of coal	-	75 per cent
Combustion efficiency of gas	-	90 per cent
Equivalent coal	-	1.33 Kg/NM^3

Therefore, 1 M^3 of distillery waste can save 33-50 kg. of coal.

Other than this saving of transportation cost of coal, saving of energy in terms of diesel and conservation of environment are indirect benefits.

Conclusions

The foregoing discussion evidently revealed that biogas is highly beneficial. It can reduce the problem of energy crisis and environmental pollution. Therefore, it is essential to popularize anaerobic treatement as an obligatory step to all, producing/ discharging organic wastes into the water bodies.

References

Gautam, Ashutosh (1992). Energy from waste and pollution control. Chemical Engineering World. 28(9): 89-90.

Kaul, S.N., Nandy, T. and Trivedy, R.K. (1995). Pollution Control in Distilleries. Enviro. Media, Karad, India.

Sastry, C.A. (1989). Water pollution control. Industrial Safety Chronical. Oct-Dec'89. 48-61.

29

Some Ecofriendly Approaches for Integrated Biomanagement of Industrial Waste Water

Manish C. Varma and Sunil K. Singh
University Department of Zoology,
T.M. Bhagalpur University,
Bhagalpur - 812 007.

Abstract

A clean environment is most important for the substance of present and future generations. Industrial waste water mainly charcaterised by high temperature, suspended solids, oils, gases, acids and certain toxic chemicals including trace metals. Such waste water causes lowering of dissolved oxygen level in receiving aquatic system, high acidity, less penetration of sun light due to oil and grease and higher assimilation of toxic chemicals. All these are posing serious threat to natural resources through point and non-point sources of contamination. For mitigating the detrimental effects certain standard physical and chemical methods have been adopted. Since all these are not adequate in minimising the toxic chemicals particularly the heavy metals, certain proposals are laid down in the present endeavour to secure clean industrial production integrated with enviornmental protection using resource management principles of eco-system. In practice eco-principles are applied to set-up appropriate technologies where environmental protection is integrated in industrial production process towards environmentally sustainable technology. In this integrated bio-system use of living machines (Phytosystem) are particularly useful in waste water management with 4-R concepts *i.e.* refuse, reuse, recycle and rehabilitation. Cyanobacterial application with macrophyte growth in lagoons prepared in the vicinity of industrial units for storage of waste water has been proved to be beneficial. Culture of molluscs in such lagoons may be practised as molluscan shells fixes heavy metals by absorption as well as adsorption.

Introduction

A clean enviornment is most important for the sustenance of present and future generations. The pace of rapid industrialisation and their subsequent discharges in the environment has created à serious threat to the naturela resources particularly the water bodies. Most of the industries are located along with large water bodies as majority of them needs plenty of water during their different stages of production. Beside taking the water for their use, it is a common practice by these industries to dump their discharges in such aquatic bodies. These practice is based on the old dictum that "better solution of pollution is dilution". These wastes commonly consists of oxygen demanding substances, chemical contaminants and toxic heavy metals. This has further worsen the situation as the load of population, eroded sediments, pollutants arising from agricultural land use, the salinization and acidification of soils, wastes from sewage treatment works, radionucleides and heated effluents from power plants, are already taking their toll, (Connell and Miller 1984). Boon (1992) remarked that today's most serious water pollution problems occur in rapidly developing countries such as India, Brazil etc. Here, some perpetrual impediments like cost management, lack of public awareness and a will to survive in international market has made the environmental management difficult. As a result very few of the large water bodies, rivers retain their original functional integrity (Karr 1993) and many have probably lost much of their capacity to adjust and recover from such disturbances.

Keeping in view the deleterious effects of industrial discharges, an attempt has been made to the substantiate the impact of effluents of Indian Copper Complex, a unit of Hindustan Copper Limited, on the riverine ecosystem of Subernarekha. This is a giant copper producing unit and releases its discharges in the nearly flowing river Subernarekha at Ghatsila (22°34'N lat. and 86°28'E longtd.)

The physical and chemical composition of this effluent is as follows:

	Effluent characteristics of HCL	
Total amount of effluent	159 m^3/day (approx) (average value of 12 months)	ISI limit
Temperature	41.5°C	40°C
pH	4.2	5.5 to 9.0

Total Suspended Solids (TSS)	1200 mg/l^{-1}	100 mg/l^{-1}
Residual Chlorine	0.35 mg/l^{-1}	1.0 mg/l^{-1}
Heavy Metals	Six metals have been identified in the effluent (average of 12 values)	
Copper	4020 ± 289 mg/l^{-1}	3 mg/l^{-1}
Zinc	2406 ± 112 mg/l^{-1}	5 mg/l^{-1}
Lead	460 ± 53 mg/l^{-1}	0.1 mg/l^{-1}
Iron	5627 ± 318 mg/l^{-1}	
Chromium	11 ± 3 mg/l^{-1}	2 mg/l^{-1}
Cadmium	12 ± 1.5 mg/l^{-1}	2 mg/l^{-1}

(Varma and Singh 1998)

The effect of industrial discharges depends upon the raw material used and the manufacturing process involved, Hodges, 1973.

The major pollutional effects of waste water on receiving water bodies

(i) Raises the termpature of the water.

(ii) Causes taste and odour problems.

(iii) Lowering of dissolved oxygen.

(iv) Deposition of suspended matter.

(v) Floating of oil and grease on water surface leading to reduction in the diffusion of oxygen and other gases inhibiting self purification of water bodies.

(vi) Coal tar as refuse choke pipes and wastes treatment appartus.

(vii) Toxicity of aquatic life.

Pollution Causative factors of Waste Water

Heat	-	Cooling water, bioler blow-down etc.
Suspended Solids	-	Sand, over burden fines, flyash coal, coke etc. Colloidal solids (mostly inorganic non-biodegradable),

		oxygen demanding heavy metals (released in different manipulations considered as most serious aquatic pollutants).
Oil and Grease	-	Rolling mill oil, lubricants, hydraulic oils, solvents etc.
Acids	-	Pickle liquor, acid salvage used mostly for cleaning the metal (pickling) or for ion-exchange.

In India promulgation of Water Act (Preventation and Control of Pollution Act, 1974) has made the industries obligatory to treat their waste before its disposal. Enforcement of industries specific standards for effluents. Minimal National Standards *i.e.* MINAS have evolved on the basis of techno-economic consideration. These results in large numbers of waste water treatment plants, but their ineffective planning, high cost of treatment processes, lack of determination and sound management on part of industries and lack of social awareness for these lapses have failed to provide the desired levels of results.

Quality standards for receiving waters are based primarily on the protection of public health, the conservation of eco-system, prevention of displeasing asthetics and sustenace of economy dependent on water for its existence.

The most simple solution which is easy to propose but impossible to implement is prohibit the discharge of pollutants in water. Since the resources are very limited, the water quality has to be attained and the benefits has to be derived, therefore, are interrelated and are of great concern.

In a metal processing industry large amount of water is required. Some part of this water is lost by evaporation and the remainder in a more or less degraded condition is returned to inland water sources. Nearly two thirds of this water used is utilised for heat exchange. It is not directly brought into the contact even with the raw materials or the finished products hence usually remains clean. Such flow, if segregated from polluted one, do not require any treatment other than cooling.

Suggestive Methods of Treatment, Utlilisation and Disposal of Effluents

The basic aim of all control measures and management is achieving and maintaing water quality at acceptable levels. In the present endeavour an effort has been made to suggest a scheme for effluent/waste water treatment from Metallic Industries. For this, mechanical, chemical and biological steps are proposed to achieve a better result.

Monitoring of effluent

In the preparation of any scheme for treatment of industrial effluents, it is essential to study the effluents flow from its sources over a 24 hrs. Period for several days to ascertain maximum, minimum and average flow. This can be achieved either by installation of flow meter or effluent holding tank provided there is any difficulty in installing the flowmeter at its source. Due care should be taken for occasional discharges due to leakage etc.

Each effluent source has to be analysed individually over 24 hour period with samples drawn regularly. The samples may be collected hourly and made into 3-6 composite samples depending on the variation of flow and composition (EPA, Manual, 1972).

Segregation of effluent

The individual effluent streams have to be segregated (mechanically) keeping in view with their volume, pollutant content, frequency of discharges etc.

The following steps should be followed, wherever applicable:

(i) Effluents containing suspended solids above 100 mg/l should be combined together as far as possible.

(ii) Oil bearing effluents should be combined.

(iii) Highly acidic and alkaline effluents should be separated from rest of the effluent streams.

(iv) Acid slurry should be separated from rest of the effluents.

(v) Sewage effluent should be treated separately as far as possible.

Sometimes the segregated effluents can be combined in such a way that it can be utilised for the treatment of others. In some cases judicious combination reduces the cost of chemicals and increases the efficiency of treatment rendered.

Treatment Methods

Treatment of metallurgical unit effluent is carried out in several steps (Sax, 1975). These steps may be combined or carried out individully depending on the nature and compassion of effluent from different metallurgical units. In different steps mechanical flocculation and clarification are to be carried out in order to prepare it in better way for chemical treatment.

Chemo-mechanical Methods

Steps

(i) *Cooling off the flow:* With proper aeration in large cooling tanks.

(ii) *Neutralisation of acids:* Effluents with low pH may be run through the carbonate boulders, lime stone sieves on tricking filter system. It may further be treated with lime after possible acid removal.

(iii) *Acid slurry seperation:* Acid should be recovered from pickle liquor to the extent possible economically. The slurry should be thickened (in slurry thickening pond) and incinerated (in incineration chamber). The ash may be used in land filling or in other industrial unit.

The sedimented slurry treated with oxygen plant waste permitted the recovery of considerable quantities of self fluxing ore fines. It has also been observed that the mixing of the oxygen plant waste and scrubbing waste permits the reuse of more than 95 per cent of water (Bonner, 1981).

(iv) *Oil and Grease seperation*: By flocculation and other suitable mechanical method, like stripping and extraction or by gravity method.

(v) *Seperation of suspended solids:* The waste water which contains high suspended solids may be collected and treated by settling (in settling tanks) and further seperation may be carried out by chemical coagulation.

Use of coagulants such as alum, ferric chloride, ferric sulphate and polyelectrolyte may be helpful not only in reducing effluent turbidity but also in compacting sludge.

(vi) *Liming:* Proper oxygenation (aeration) of treated waste water coming from different sources should be assured before final treatment with lime. Steps should also be taken to reclaim the mixed lime from lime-slush thickening tank by its proper incineration. Lime treatment or alkalisation also eliminates higher release of metals etc. in the form of suspended solids.

(vii) *Phosphatization of metals:* Super phosphate eliminates the higher release of suspended metals without coagulating it.

(viii) *Clarification:* Clarification pond (sludge accumulator) reduces the concentration of suspended solids including different metals subsequently.

(ix) *Crystallisation of inorganic solubles:* This may be achieved by the use of various chemical coagulants such as aluminium sulphate, ferric chloride or ferric sulphate (pickle liquor along with sacked lime). The dry cake can be disposed off as a land fill.

Phytotreatment or Biological Treatment

Effluent from chemo-mechanical regeneration system should be pumped into the bottom of 4′ to 5′ deep stabilisation ponds pre-cultured with mixed phytopopulation. Such ponds operate in a aerobic anaerobic system in such a manner as naturally fertile lagoons.

Algal Treatment

Cyanobacteria have been found useful with a variety of effluents because they interact with pollutants and metabolize or use them as nutrients. They detoxify or degrade them and accumulate or absorb them to their cells (Subramanian and Uma, 1996).

The benefits of algae in treatment systems are oxigenation and mineralization in addition to their role as producers in the trophic eco-systems (Forstner U and Whittman CTW, 1981). The luxuriant growth of algal forms in euthrophic waters is a common phenomenon. The role of algae in the removal of various kinds of inorganic and related substances by their metabolic activities has been studied by several workers (Benemann *et al.*, 1977; Gupta and

Prabhakar, 1980; Nair *et al.*, 1981; Senger *et al.*, 1985; Senger and Sharma, 1987; Senger *et al.*, 1990). Algae being the primary producers and ubiqoutos in nature almost serves as indicator of habitat condition (Palmer, 1980; Elnsabarawy and Welter, 1984; Gunale, 1991; Varma *et al.*, 1991; Singh and Varma, 1999).

Algal bacterial symbiosis has long been proved to be an inexpensive process for reclamation of waste water (Gotass *et al.*, 1951; Gotass and Oswald, 1955; Ganpathi and Amin, 1972; Dox and Svi, 1980; Subramania and Shanmugasundram, 1986). The algal system particularly the cyanobacteria is ideally suited in treating the waste water due to its high flexibility to adopt in various environment (Walsh and Merril, 1984). It also produces a variety of useful by-products with known nutritional and fertilizer values from their biomass (Subramanian and Shanmugasundram, 1986). Employment of myxohyceae population in contaminated water may prove beneficial in many ways. They bring about oxygenation in water in additon to being a good food source for aquatic species (Elnsabarawy and Welter, 1984).

Uma and Subramanian, 1990 have studied the effective use of myxophyceae in industrial effluent with a very high level of total dissolved solids (TDS), calcium and chloride content and which cannot be treated successfully by the conventional treatment system consisting of settlement tanks and aerobic digester. They have used a halophilic bacterium. *Holobacterium* sp. 101, a marine filamentious Myxophyceae *Oscillatoria* sp. BDU101 42 and a calcicolous unicellular Myxophyceae *Aphanocastra* sp. BDU 16 both in laboratory as well as in field conditions. With a serial inoculation of these three organisms for a retention time of four days, it was rewarded more than 50 per cent reduction of chlorides and nearly 50 per cent reduction in calcium in laboratory condition, in field conditions this was 40 per cent and 25 per cent respectively, which is still more efficient than conventional treatment methods.

Reports are available on the reduction of BOD, COD and high level of nitrogenous compounds by algal culture in different industrial waste water (Saxena *et al.*, 1974; Govindan, 1985; Tam and Wong, 1989). The capacity of Myxophyceae to remove large amount of phosphorous from industrial waste water has been demonstrated by several workes (Stevenson, 1967; Chan *et al.*, 1979; Tam and Wong, 1989). They observed the release of organic acids by microbes and especially Myxophyceae and their capacity to solubulize magnesium in the waste water.

Mittal *et al.*, 1992, have studied the toxicity and uptake of some heavey metals like Fe, Zn and Cu by *Oscillatoria prormata var. attenuata* and *Scenedesmus quadricauda var. longispina*. They found that concentration up to 20.67; 0.595 and 0.0188 mg ml^{-1} for *Scenedesmus* and 27.56; 0.833 and 0.0207 mg mg^{-1} for *Oscillatoria* of Iron, Zinc and Copper respectively in the medium favoured growth of algae and later on their reduction.

The capacity of algae to reduce the heavy metal pollution load in river water has been tested by Senger *et al.*, 1990 by culture experiments. They found that on 30th days of growth there is a complete removal of Fe, Zn and Cu by mixed algal culture. *Chlamydomonas confera* removed Fe^{++}, Cu^{++} completely while *Phormidium conium* removed Zn^{++} and Cu^{++} completely on 30th days.

Sallal (1982), found that the heavy metals such as Fe, Zn, Mn, Pb, Ni, Cd, Co, Cr, etc. favoured the growth of algae and blue green algae and their successive absorbence in feed back mechanism. All these workers suggest for a better result with the biological treatment of the pre chemo-mechnical treated effluent in pre-cultured shallow tanks.

Rai and Chandra (1989), experimented and concluded that *Hydrodictyon reticulatum* (water *net al*gae) accumulates Cu, Pb, Mn and Fe both under field and laboratary conditions. Similarly results have been shown by *Chara corallina.* The potential of both these algae to remove metals in preferential sequence of Fe>Mn>Pb>Cr>Cu>Cd and are able to reduce the level of some metals permissible level.

Role of macrophytes in pollution abatement

Aquatic macrophytes (free floating, submerged, emergent) are ideally suited for remeadication of industrial waste water. Mixed use of species belonging to the genra *Spirodela, Lemna, Ceratophyllum, Eichhorina, Jussiaea, Ipomea, Typha* etc. may exploit all the zones of lagoon. Free floating macrophytes particularly vascular one has to be proved effective.

Wolverton and McDonald (1970), suggest the use of highly productive water hyacinth as an execllent aqua purifier. They imbibe certain toxic heavy metals, under the laboratory condition. It has been proved that removal of 0.67 mg of cadmium, 0.5 mg of nickel and 0.1 mg of mercury per gm dry weight of water hyacinth possible in 24 hrs. Sometimes its excessive growth lead to depletion of dissolve oxygen content of the water body which affect the biodiversity of the system. Duck weed (*Spirodela polyrhiza*) has been found effective in removing Cd, Cr, Fe, Mn (Rai and Chandra, 1992). They are quite safe and can be harvested easily. *Pistia stratiotes* can remove appreciable amount of Cd (Rai *et al.*, 1995). Submerged plants complete their life cycle inside water and have more capability to absorb pollutants [Sinha *et al.*, 1994(a)]. This property is mainly due to their higher surface/biomass ratio than floating and emergent plants. *Certophyllum demersum* and *Hydrilla verticillata* have shown excellent results (Sinha and Chandra, 1990, Sinha *et al.*, 1993). But they grow only in oxygenated water hence where microbial decomposition creates anoxic decomposition they can not grow properly as well as their harvesting is also difficult.

Emergent and partially emergent plants with hollow stems and well developed root zones are excellent aquapurifier. They grow luxuriantly in shallow water bodies and particularly they are very efficient in removing metals from sediments (Rai *et al.*, 1995). It has been found that the accumulation of metals in the root is more marked than the leaves. They are useful in bio-monitoring too as they do not migrate and attain equilibrium with their surroundings. Leaves, stems and roots of these plants are very useful as they can be used for making paper and pulp in view of their high fibre contents. *Scirpus lacustris* partially emergent plant effective in removing Cu and Cr because of their configuration, the leaves being tall, cylindrical, hollow and facilitating transport of atemospheric oxygen to the rizophere zone. *Bacopa monnieri* (rooted, partially emergent) a common wet land plant has shown ability to accumulate Cu and Cd from the polluted waste water (Sinha and Chandra, 1990). *Vallisneria spiralis* removes high amount of Pb, Cu, Cd [Sinha *et al.*, 1994(a)]. Similarly use of Cyperus sp. *Eichorrena crassipes* and *Typha latifolia* have been proved very effective in removing Cu, Fe and Zn (Rai *et al.*, 1995).

Hence application of biological methods in treating industrial effluents would be of great advantage.

The advantage of biological waste water treatment system

(i) Low operative cost + to ensure clean industrial production.

(ii) They can often establish at rites where the waste water is produced.

(iii) They are more flexible and less susceptible to shock loading.

(iv) On cost management account it is well suited to third world countries.

(v) Ecological diversity is sustained at all levels by co-existence of living being including man in harmony with water embeded in land system.

The potential advantages of Algae in treating industrial waste water

The luxuriant growth of algal forms in eutrophic waters is a common phenomenon:

(i) Their role in the removal of various kinds of inorganic and related substances by their, metabolic activities has been established by several workers.

(ii) Several workers have established the algal bacterial symbiosis in an inexpensive process for reclamation of waste water.

(iii) The algal system particularly the Myxophyceae is ideally suited to treat the waste water due to its high flexibility to adopt in various environment.

(iv) They bring about oxygenation in water even at diffused sun light.

(v) Act as good food source for aquatic species.

(vi) Produce a variety of useful bioproducts with known nutritional and fertilizer values from their biomass.

The potential of Macrophytes in waste water treatment system

(i) Macrophytes when utilised in a controlled biological system constitute a remarkably efficient and inexpensive filtration and disposal system for toxic metals.

(ii) Macorphytes inherent potential to concentrate metal from their surrounding water and sediments in which they grow.

(iii) They help in pollution assessment. Low levels of pollutants can also be easily detected as plants bio concentrate the metal in their tissues to serve as warning system in pollution control activities.

(iv) In combination, they reduce the levels of metals in polluted water below the maximum permissible limits.

Constraints

(i) In some location for such a purpose may prohibitively expensive in using such treatment techniques.

(ii) There is a potential hazard from pathogens, toxic heavy metals and carcinogens. Thus, public health safety measures should be taken care of.

Algal / Macrophyte	*Role*
Holobacterium sp. 101	Serial inoculation of these three with four days retention time removes 50 per cent of Chlorides and Calcium under lab. Condition. 40 per cent and 25 per cent under field condition.
(Holophilic bacterium)	
Oscillatoria sp. BDU 10142	
(Filamentous myxophyceae)	
Aphanocastra sp. BDU 16	
(Calcicolous unicellular myxophyceae)	
(Uma and Subramanian, 1990)	
Oscillatoria prornata var. Attinuta	Fe 27.56; Zn 0.833, Cu 0.0207 μ g ml^{-1} in the medium.
Scenedesmus quadricauda var. longispina (Mittal *et al.*, 1992)	Fe 20.67; Zn 0.595; Cu 0.0188 μ g ml^{-1} in the medium favoured the growth of these two and later their successive reduction.
Chalmydomonas confera	Mixed culture of these two remove Fe, Zn, Cu completely on 30th day of its culture under lab. Conditions.

Phormidium conium (Senger *et al.*, 1990)	
Hydrodiction reticulatum (Rai and Chandra, 1989)	removes Cu, Pb, Mn, Fe both under field and lab. Conditions.
Chara corallina (Rai and Chandra, 1989)	removes Cu, Pb, Mn, Fe both under field and lab. Conditions.
Water hyacinth (Wolverton and McDonald, 1970)	removes Cd 0.67 mg, Ni 0.5 mg, Hg 0.1 mg/gr dry weight of water hyacinth in 24 hrs. under lab. Conditions.
Duck weed (*Spirodela polyrhiza*) (Sinha *et al.*, 1994a)	removes Cd, Cr, Fe, Mn
Pistia stratiotes (Rai *et al.*, 1995)	removes Cd
Ceratophyllum demesrsum *Hydrilla verticillata* (Sinha and Chandra, 1990)	removes Pb, Ni, Cu
Limnantheum cristatum; Wolfia globosa (Rai *et al.*, 1995)	removes Cd, Cr
Scirpus lacustaris; Bacopa monnieri (Sinha and Chandra, 1990)	removes Cd, Cd, Cr
Vallisneria spiralis	high amount of Pb
Cyperus sp.; E. Crassipes, Typhalatifolia (Sinha *et al.*, 1994b)	All these three are effctive in removing Cu, Fe, and Zn

The entire process may be summarised as follows

Process	Methods
1 Maintaining and Monitoring the flow of effluent	Minimum and maximum flow of effluent over 24 hrs. Period to be measured either by flow meter or measuring tanks on hourly basis.
2 Segregation of effluents Mechanical	(a) Cooling of effluents (b) Suspended solids above 100 mg/L^{-1} be combined together as far as possible. (c) Acidic and alkaline effluents seperated as far as possible. (d) Acid slurry separation (separation of possible acids by pickle liquor method) (e) Separation of oil and grease.

Process	*Methods*
Chemical treatment	(a) Neutralisation of acids by liming or Phosphatisation (b) Crystallisation of inorganic soluble (c) Coagulation of suspended solids including metals (d) Oxygeneration aeration of effleunts (e) Liming (f) Flocculation (g) Phosphatisation of metals (h) Clarification for removal of additional phosphate
Biological treatment	Effluent from different individual treatment plants comes to a common mixing plant, from there through trickling filter plant these are settled in shallow stabilising tanks (4' to 5' deep) for Phytoremedication of heavy metals. The surface of the trickling filter can be further reinforced by suitable microbial decomposition system in order to eliminate some toxic metals.

Conclusion

The task of saving enviornment may go beyond unending discussions but we will have to realise the importance of this vital issue. State has enforced many stringent regulation and control to overcome degradation in further course but without voluntary efforts of individuals, out come "remians drop in the ocean". Development is required to meet our everyday's needs, but definitely not at the cost of environment. Today we indeed are passing through a phase of transition and, thus, a lot is to be done. So, we might have to (though nobody wish to) encroach the territory of environment but the only thing to be taken care of is to be within the acceptable limits of our mother earth.

No water body receiving effluents can be kept as clean as virgin streams. Still, effort should be made to keep them clean enough to that the down stream users do not have to resort to another remedial measures or become target of all miseries of industrialisation.

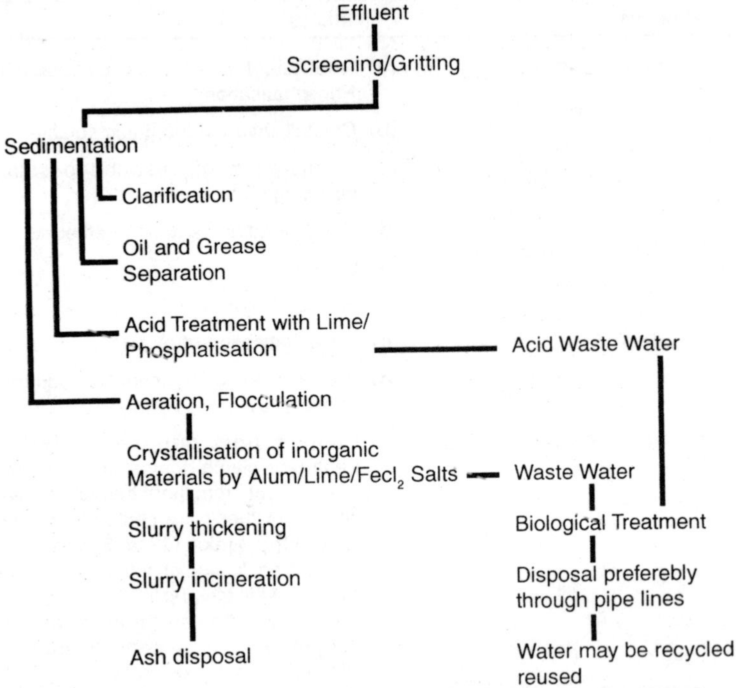

Chart: Flow sheet depicting different stages suggested for effluent treatment

Further Suggestions

(i) If specific locations and receiving enviornmental system so demand, discharge limitations should be based on local receiving water standards with taking into account the self-purifying capacity of natural water bodies.

(ii) Regular water quality and quantity survey with its physical, chemical and biological phenomenon should be done.

(iii) While treating the effluent, preference should always be given to such schemes where recovery of waste products for reuse can be made.

(iv) The effluent water after adequate treatment can be recycled. This reduces water consumption as well as the final volume of discharge.

(v) The industry should be able to discharge effluent of such a character that the treated water can be used within the plant *i.e.* the ability of the plant to recycle the effluent into its own service.

(vi) The effluent character should be such that the desired quality of the receiving water is attainable with a minimum dilution.

(vii) Finally, letting nature to do a lot of work the areas of land and water have to be set aside where semi natural ecosystem have to be developed. New large industries should be zoned and sited in the middle of natural areas large enough for treatment of degradable waste and containment of poisonous water so they should never find their way into general environment.

Disposal

The final disposal of the treated effluents beyond the factory boundary is an important step. It is preferable to give sufficient detention time in stabilising ponds. The overflow from these stabilising ponds passes to the effluent drain leading to the receiving natural system. The treated effluent waste water in the effluent drain must be in conformity with the MINAS. In case the drain passes through a locality where there is a possibility of use of this water as raw water source by the inhabitants and cattle, suitable protection of the drain from the approach of the people and cattle with proper warnings to be made. In some cases, it is preferable to discharge the treated effluents through a pipeline to the receiving water. When all the characteristics of the individual effluent streams of the process plant are properly assessed the effluents discharged from effluents treatment units can also be evaluated with respect to the extent of treatment plants.

The final effluent characteristics can be predicted and made to comply with the requirements prescribed by the regulatory authorities.

However, the treatment and purification process is never hundred per cent complete and "purified" industrial waste water cannot be assumed to be fit for drinking purposes.

References

Benemann, J.R., Koopman, B.L., Becker, D.C., Weisman, J.C. and Oswald, W.J. (1977). A systems analysis of bioconversion with micro algae. Paper present at symp. on clean fuels from biomass and wastes. 25-28. Orlando, Florida.

Bonner, T. (1981). Hazardous Incineration Engineering, Pollution Technology Review, No. 88, Noyes Data Corporation, Park Ridge, N.Y.

Boon, P.J. (1992). Essential elements in the case for river conservation. In: P.J. Boon, P. Calow and G.E. Petts, editors. River Conservation and management. John Wiley and Sons, Chichester, pp. 11-33.

Chan, K.Y., Wong, K.H. and Wong, P.K. (1979). Nitrogen and Phosphorus removal from sewage effluents with high salinity Chlorella salina Environ. Pollut. 18: 139-146.

Connella, D.W. and G.J. Miller (1984). Chemistry and ecotoxicology of pollution. John Wiley and Sons. New York.

Dash, A.K. and Mishra, P.C. (1998). Role of Cynobacteria in water pollution abatement, Indian J. Env. and Ecoplan. 1(1 and 2): 1-11.

Dox, I. and Svi, B. (1980). Effect of hetrotrophic bacterias on green algae growing in waste water. In: Algal Biomass (Eds. G. Sheief and C.J. Soeder). Elsevier/ North Holland Biomedical Press. pp. 421-429.

Elnsabarawy, M.T. and Weller, A.N. (1984). Utilization of algal cultures and assays by industry. In: Algae as Ecological Indicators (Ed. L.E. Shubert). Academic Press Inc., pp. 317-328.

EPA (1972). Manual of Test Procedures, method 1110; adapted from National Association of Corrosion Engineers Standard TM. 01-69.

Forstner, U. and Whittman, C.T.W. (1981). Metal pollution in the aquatic environment. Springer-Verlag, Berlin.

Ganapathil S.V, and Amin, P.M. (1972). Studies on algal bacterial symbiosis in low cost water treatment systems. In: Taxonomy and Biology of BGA (Ed. T.V. Desikachary) pp. 483-493.

Gotass, H.B., Oswald, W.H., Ludwig, H.E. and Lynch, V. (1951). Algal Symbiosis in oxidation ponds. Second progress reprot. University of California Inst. Eng. Res. Bull. 44. December.

Gotass, H.B., Oswald, W.H. (1955). Transaction of the conference of the use of solar energy. The Scientific Basis Tuscon, Arizona. 4: 95-114.

Govindan, V.S. (1985). Treatment of tannery waste water by stabilization pond methods. Indian J. Environ. Health. 27: 58-66.

Gupta, S.K. and Prabhakar Rao, V.S.S. (1980). Treatment of urea by algae, activated sludge and flocculating algal bacterial system - A comaprative study. Indian J. Enviorn.Hlth. 22: 103-112.

Gunale, V.R. (1991). Algal Communities as indicators of Pollution. J. Environ. Biol. 12: 223-232.

Karr, J.R. (1993). Protecting ecological integrity: An urgent societal goal. Yale Journal of International Law 18(1): 297-306.

Hodges, L. (1973). Environmental Pollution, Holt. Rinoholt and Winston, Inc. New York.

Manoharan, C. and Subramanian, G. (1992b). Sewage-Cynobacteria interaction: A case study. Indian J. Enviorn. Protect. 12(4): 251-258.

Mittal, S., Sengar, R.M.S. and Kaushik, B.D. (1992). Uptake and toxicity of heavy metals to algae. Indian J. Microbiol. 32(1): 51-53.

Nair, K.V.K., Kannan, V. and Sebastian, S. (1981). Nutrient removal from sewage effluents using Scendesmus. Indian J. Environ. Hlth. 23: 53-59.

Palmer, C.M. (1980). Alage and water pollution. Castle House Publication, England.

Rai, U.N. and Chandra, P. (1989). Removal of heavy metals from polluted by *Hydrodictyon reticulatum* (Linn.) Lagerheim. Sci. Total environ. 87/88, 509-515.

Rai, U.N., Tripathi,R.D., Gupta, M. and Chandra, P. (1995). Introduction of Phytochelatins under Cadmium stress in water lettuce (*Pistia startiotes*) J. Environ. Sci. Health A 30(9): 2007-2026.

Rai, U.N. and P. Chandra (1992). Accumulation of Copper, Lead Magnese and Iron by field population of *Hydrodictyon reticulatum* (linn.) Lagerheim. Sci. Total Environ. 116: 203-211.

Rai, U.N., Tripathi, R.D., Sinha, S. and Chandra, P. (1995). Chromium and Cadmium bioaccumulation and toxicity in *Hydrilla verticillata* (l.f.) Royle and Chara Corallina Wildenow J. Envi. Sci. Health. Part A 30(3): 537-551.

Rai, U.N., Sinha, S., Tripathi, R.D. and Chandra, P. (1995). Waste water treatibility potential of some aquatic macrophytes. Removal of heavy metals. Ecological Engineering (5): 5-12.

Saxena, P.N., Tiwari, A. and Khan, M.A. (1974). Effect of Anacystis nidulans on physiochemical and biological characteristics of raw sewage. Proc. Indian Acad. Sci. 79: 139-146.

Sax, N.I. (1975). Dangerous properties of industrial materials, 4th edn. Van Nostrand Reinhold, New York.

Sallal, A.K.J. (1982). Growth of algae and cynobacteria on sewage effluents. Microbios Letters. 20: 7-14.

Senger, R.M.S., Sharma, K.D. and Pathak, P.D. (1985). Studies on distribution of algal flora in polluted and non-polluted region in Yamuna river at Agra (U.P.). J. Indian Bot. Soc. 64: 365-376.

Senger, R.M.S. and Sharma, K.D. (1987). Tolerance of *Pharmidium corium* and *Chlamydomonas* Sp. against chemical elements present in polluted water of Yamuna river. Proc. Nat. Acad. Sci. India. 57: 56-64.

Senger, R.M.S., Sharma, K.D. and Mittal, S. (1990). In vitro studies for the purification of river water by algal treatment. Geobios. 17: 77-81.

Singh, S.K. and Varma, M.C/. (1999). Biomonitoring of river Subernarekha polluted with industrial discharges. Environment and Ecology 17(1): 79-87.

Sinha, S. (1998). Final technical report on "Identification of suitable plant species for Biological water purification." Dept. of Rural Dev. Govt. of India.

Sinha, S. and Chandra, P. (1990). Removal of Cu and Cd from water by *Bacopa monnieri*. Water, Air and Soil Pollution. 51: 271-276.

Sinha, S., Rai, U.N., Tripathi, R.D. and Chandra, P. (1993). Chromium and manganese Uptake by *Hydrilla verticillala* (l.f.) Royle: Amelioration of Chromium Toxicity by manganese. J. Environ. Sci. and Health. A28(7): 1545-1552.

Sinha, S., Rai, U.N. and Chandra, P. (1994a). Accumulation and toxicity of iron and manganese in *Spirodela polyrrhiza* L. (Scheilden). S. Bull. Environ. Coutam. Toxicol. 53(4): 610-617.

Sinha, S., Gupta, M. and Chandra, P. (1994b). Bioaccumulation and toxicity of Cu and Cd in *Vallisneria spiralis* (L). Envlron. Monito. Assess. 33: 75-84.

Stevenson, F. J. (1967). Organic acids in Soil. Soil biochemistry (Eds. A.D. McLaren and G.R. Peterson) Marcel Dekker, New York, pp. 119-146.

Subramanian, G. and Shanugasundram, S. (1986). Sewage utilization and waste recycling by cynobacteria. Indian J. Environ. Hlth. 28: 250-253.

Subramanian, G. and Uma, L. (1996). Cyanobacteria in pollution control, Journal of Scientific and Industrial Reserach, Vol. 55, Aug-Sep., pp. 685-692.

Tam. N.F.Y. and Wong, Y.S. (1989). Waste water nutrient removal by *Chlorella pyrenoidosa* and *Scenedesmus* Sp. Enviorn. Pollut. 58: 18-34.

Uma, L. and Subramanain, G. (1990). Effective use of cynobacteria in effluent treatment. Proc. Nat. Symp. on Cyanobacteria in nitrogen fixation IARI. New Delhi. pp. 437-443.

Varma, M.C., Singh, N.K. and Munshi, J.S.D. (1991). Algal Species diversity as an index of water quality of river Subernarekha at Ghatsila. J. Freshwater Biol. 3: 235-239.

Varma, M.C., Singh, S.K. (1998). Industrial discharges and its chemical specification in the river Subernarekha, Journal of Environment and Pollution 5(4), pp. 317-322.

Walsh, G.E. and Merrill, R.G. (1984). Algal bioassays on industrial and energy process effluents. In: Algae as Ecological Indicators (Ed. L.E. Shubert). Academic Press, London. pp. 330.

30

Rotifera : Interaction with the Atmosphere

S. Kumar, P. Kumari, Subranodeo and Rajnish Kamal
Department of Zoology,
P.Sc. College,
Madhepura, Bihar

Introduction

The study of ecobiology has emerged as a major biological discipline in recent years. A number of workers have made extensive researches on the ecobiological studies of certain dominant micro and macrobiota in both terrestrial and aquatic habitats of India and abroad. The present investigation has been taken up to assess the rotifera and their interaction with the environmental condition in two fresh water fisheries ponds (Namely - Pond A and Pond B) of Kosi area (Madhepura) Bihar. The city of Madhepura is located near the river Kosi with North latitude of 25°8′ to 26°20′ and 86°50′ to 87°9′ East longitude and area $(1788.5)^2$ Km.

North Bihar especially the Katihar, Purnea, Madhepura and Saharsa districts are maximum effected areas of expected flood of Kosi-river spreads over more than 35000 sq km areas and gets detached during dry seasons from the main current. Such in undated water areas represents a marvellously productive fresh water habitats with a diverse floristic and faunistic composition like copepods, cladocerans, Rhizopods including zooplanktonic rotifer fauna which are important item of fish food and are immense help in piscicultural practices and shows numerical superiority over other groups in different environmental conditions of different seasons of the year.

Material and Methods

Plankton hauls were made regularly from different sectors of the pond at a particular time with the help of a plankton net of mesh size 25 cm and the different zooplankton species and phytoplankton collected were preserved in 5 per cent formalin and studied under the research microscope. Further, from collecting biomass different types of food content of rotifers and diverse group of rotifers species were separated, studied and identified as per description given in Ward and Whipple (1959), Needham and Needham (1972), simultaneously, these water samples were also analysed for the determination of temperature, dissolve oxygen contents and rainfall data of the study area. The temperature was read on thermometer at every point with the water column. It was measured with the help of a Mercury Themometer graduated upto 110°C. Dissolve oxygen content analyzed by Winkler's method and APHA (1971). Rainfall data also observed and noted regularly from Agricultrual Station, Madhepura.

Transfer of solar energy to exist the rotifer population carried out by standard method adopted by Gliwicz (1969). In order to study, the phenological aspects of certain dominant species of rotifera of the present investigation, 5 sets, each containing 10 samples of a particular identified female rotifer sepcies are placed on a petridish at room temperature of the laboratory. Water sample of the same pond containing certain algal matters like *chlorella, scendesmus* etc. were placed on the same petridish. Thus, the rotifers were fed with these algal matters and allowed to grow. Egg development and growth of these rotifers were noticed from time to time.

Sometimes, culture media was also prepared in the laboratory and identified rotifer species were allowed to grow.

Observations and Result

It is evident from the present investigation that the diverse group of rotifera species have a great iteraction with the environment proper and different environmental factors. The occurrence, abundance, density, seasonality and periodicity including the phenomenon of cyclomorphosis and apomictic parthenogenesis of rotifer is purely depending upon physico-chemical conditions of the medium and rainfall data of the period. Cyclomorphosis is the phenomenon of changes in the body form and structure of an organism in different seasons of the year. This phenomenon is also exhibited by other zooplankton species of cladocerans also as is reported by Brooks (1946). This cyclomorpyhosis found in the present investigation also tallies with the findings of Vasist (1977), Nayer (1965), Sharma (1976), King (1977), Lynch (1984). Apomictic parthenogenesis also gives rise to different types of genotype and for their production of eggs depends on the favourable or unfavourable condition of the environment. The different rotifer species were found in high number from April to September 98 and in May to September 99 in these ponds of Madhepura when the temperature of the medium ranged from 30.4°C to 35.4°C, Oxygen content in the range of 4.2 ppm to 14.7 ppm and rainfall data of 13.0 mm to 488.7 mm and 76.3 mm to 372.4 mm (Table 30.2). The occurrence of numerous macro and microbiota in the medium also determines the occurrence, seasonality, abundance of different rotifer species. The presence of specific macrophytes seems to be a prime factor controlling the diversity of zooplankton community. The characteristics weeds of these ponds and specially their monthly growth performance had negative significant effects on species diversity of different types of zooplankton and rotifers. Some workers like Munro (1974) and Voights (1976) held their views that greater diversity of zooplankton and rotifers exist, when different cover vegetation are interspersed. The thick vegetation, due to their manifold effects on water, such as cesation of photosynthetic activities and phytoplankton growth, cause shortage of both respiratory gases and nutrient sources and hence hamper the growth of diverse zooplankton community. On the contrary the presence of submerged macrophytes was found to accelerate significantly the specific diversity of different rotifers in pond B.

Rotifera found a very important part and link in the food chain of an aquatic ecosystem and regarded as primary consumers feeding upon a number of phytoplanktonic algae, detritus and different types of bacteria of the medium. These rotifers are found to take a particular type of phytoplanktonic algae a food in the present investigations. These food of rotifers are found to differ from those taken by other types of zooplankton like cladocerans and copepods. This fact was also confirmed by the finding of Munshi *et al.*, (1990).

The pond A of the study area contains the number of phytoplankton and macrophytic species shown in the (Table 30.1).

Hence, it is seen that the pond A contains a dense growth of both micro and macrobiota. Therefore, it is much productive in nature.

These above mentioned micro and macrobiota has got a great influence upon the abiotic parameters of the medium and also as the abundance, distribution and specially on the rotifer fauna of the ponds of Madhepura during the period of investigation (April'98 to March'99).

Table 30.1

Phytoplankton	
1	Anabaena beckl
2	Oscillatoria Limosa
3	Microcystis aeruginosa
4	Nostoc sp.
5	Gleopeapsa sp.
6	Aphanocapsa spongelli
7	Spirogyra neglecta
8	Scendesmus armatus
9	Pediastrum duplex
10	Oedogonium oblongum
11	Volvox globater
12	Cosmarium botrytis
13	Closterium sp.
14	Pondorina sp.
15	Navicula cuspidata
16	Synedra ulna
17	Cymbella turgida
Macrophyte Species (Free, Floating and Submerged)	
1	Azolla pinnata
-2	Junssinea repens
3	Lemna minor
4	Spirodela polyrhiza
5	Hydrilla verticilleta
6	Potamogeton Crispus
7	Ceratophyllum domersum
8	Euryale ferox
9	Ranunculus sclerotus

Table 30.2: Dissolve Oxygen content and materological data of the study area

Name of the month		*DO_2 (ppm)*	*Temperature °C*		*Rainfall (mm)*
			Minimum	*Maximum*	
April	98	8.8	17.1	34.4	49.9
May	98	12.2	22.1	33.8	13.0
June	98	11.4	25.5	35.4	152.4
July	98	11.8	25.1	33.0	488.7
August	98	14.2	25.5	32.6	313.4
Sept	98	8.6	25.1	32.5	76.1
Oct	98	10.4	22.6	32.8	70.0
Nov	98	6.8	16.1	30.6	3.4
Dec	98	4.8	10.5	26.7	3.7
Jan	99	4.2	8.3	23.2	22.2
Feb	99	6.4	8.2	24.4	5.3
March	99	8.2	11.3	28.4	4.0
April	99	8.6	15.8	35.4	0.0
May	99	11.4	23.4	35.4	76.3
June	99	10.6	23.9	34.9	137.4
July	99	12.4	26.1	35.4	372.4
August	99	14.7	26.1	32.4	166.2
Sept	99	5.8	26.0	32.6	345.5
Oct	99	12.8	24.3	31.8	83.9
Nov	99	7.4	17.3	30.3	0.0
Dec	99	4.2	10.9	26.1	6.7

These observation have been confirmed by the findings of Rai and Datta Munshi (1979). Further the rotifera population in pond A along with other types of micro and macrobiota was always found to dominant over other zooplankton communities in that pond. Hence, the pond A was found to be more eutropic in nature than that of pond B, as the term suggested by George (1966) and Arora (1966). Further the abundance of rotifera in such water bodies can improve the water quantity and this population can check the water body from pollution effects. This findings was also confirmed by the observation of Lund (1969).

Rotifers also play an important role on ecological bioenergetics in the environment and in the medium. It involves the process of trophic transfer and transformation of energy throuh rotifer population and community which tallies with the findings of Hubbel (1971) and Odum (1968), Haniffa and Pandian (1978). The zooplankton population specially the rotifers consume a significant proportion of the producers and the eneregy accumulation was at the rate of 0.1349 K.cal M^{-2} day^{-1}.

The phytozooplankton population linkage specially the rotifer population linkage has found to have a very low energy transfer efficiency in water bodies Kosi-division (particularly Madhepura) in the present investigation. This part of the finding also is confirmed with the findings of Gliwicz (1969) that eutrophic lakes characterisctically show low energy transfer efficiency, between primary producers and herbivore levels. Gulati (1975) and Conveny *et al.*, (1977) - have suggested that the low transfer efficiency of 4-10 per cent is usual in fresh water medium and fresh water ecosystem.

It is also clear that the rotifer population exists in all the three lines of transfer of solar energy to fish biomass in the aquatic ecosystem having different transfer efficiencies:

(1) Solar energy —0.12%— Phytoplankton : → 14.0% Catla; → 10.1 %Labeo; → 3.1% Cirrhina mirgala

(2) Solar energy —0.12%— Phytoplankton : 5.1% Zooplankton → 85.2% Catla; 7.4% Rohu

-(3) Solar energy —0.12 per cent— Phytoplankton : 5.1% Zooplankton Detritus fishes

It is confirmed with the findings of Verma (1981). Certain species of rotifers, *viz. Filinia, Brachionus* and *Keratella* are indicators of organic pollution. Environments receiving urban sewages are rich with these species due to its high tolerance capacity in response to chemical and other types of pollutants as is confirmed with the findings of Lund (1969), Stewart (1976).

Acknowledgement

Authors are grateful to Dr. S.C. Yadav, Professor of Zoology, A.N. College, Patna, for his constructive criticism and also extending our thanks to Mrs. Meera Devi and Miss Shalini whose uncalled for interference in our work added relish to our mental rigour during this trying period.

References

Arora, N.C. (1966). Rotifers as indicators of tropical nature of environment. Hydrobiol. 27 (1-2): 146-159.

Brooks, J.L. (1946). Cyclomorphosis in Daphnia, Ecol. Monograph, 16: 409-447.

Conveney, M.F., G. Enol, M. Larson, K. and Olofsson, L. (1977). Phytoplankton, Zooplankton and bacteria - Standing crop and production relationships in a eutrophic lake, Oikos, 29: 5-21.

George, M.G. (1966). Comparative plankton ecology of five fish tanks in Delhi, India. Hydrobiol. 27: 81-108.

Gliwicz, Z.M. (1969). Studies on the feeding of pelagic zooplankton in lakes with varying trophi Ekol. Polska, Ser. AA, 17: 663-708.

Gulati, R.D. (1975). A study on the role of herbivorous zooplankton community as primary consumers of phytoplankton in Dutch lakes, Verh. Int. Ver. Limnolol. 19: 1202-1210.

Haniffa, M.A. and Pandian, T.J. (1978). Morphometry primary production and energy flow in a tropical pond. Hydrobiologia. Vol. 59, 1: 23-48.

Hubbell, S.P., (1971). Of sowbugs and Systems: The ecological bio-energetics of a terrestrial isopod. In: Systems Analysis and stimualtion in Ecoplogy. Vo. 1. Acad. Press, Inc., N.Y. P. 269-324.

King, C.E. (1977). Genetics and reproduction, Variations and Adoptation in rotifers, Arch. Hydrobiol. 1 scin., B. 187-201.

Lund, J.W.G. (1969). Phytoplankton: In proceeding of the Intern. Symposium in Europe Eutropication cases, consequences conneticng - National Academy of Science, Washington, D.C., 306-330.

Lynch, M. (1984). The genetic structure of a *cyclical parthenogene* - Evolution, 38: 186-203.

Munshi, J.S.D., D.N. Singh and D.K. Singh (1990). Food and Feeding relationship of certain aquatic animals in the Ganga ecosystem. Tropical Ecology 31(2): 138-144.

Munro, I.G. (1974). The effect of temperature on the development of egg. nauphiar and copepodite Stages of two species of copepods, cyclopods, Nicines, Uljanin and Eudiaptomus gracillis, Sars. Cecolyai, 16, 355-367.

Nayar, C.K.G. (1965). Cyclomorphosis of Brachionus Calcyflorus Pallas, Hydrobiologia, Vol. 25: 538-544.

Odum, E.P. (1968). Energy flow in ecosystem: A historical review. Amar. Zoo. - 48: 11-18.

Rai, D.N. and Datta Munshi, J.S. (1979). Influence of thick floating vegetation, (water hyacinth Eichhornia crassipes in the physico-chemical Environment of a fresh-water wet land. Hydrobiologia, Vol. 62, 1, 65-69.

Sharma, B.K. (1976). Observations on a planktonic rotifer-keratella tropica (Apsteins) - Rewal, Zool. Survey, India - 2: 17-19.

Stewart, K.M. (1976). Oxygen deficite Clerity and eutropication in Some Naderan lakes. Inst. Revne. ges. Hydrobiol. 61(3): 563-579.

Vasist, H.S. (1977). Seasonal abundance of Brachionus in relation to temperature and pH. Indi. J. Ecol. 4(2): 233-235.

Verma, P.K. (1981). Limnological Survey of Badua Reservoir, Bhagalpur (Bihar).

Voight, D.K. (1976). Aquatic invertebrate abundance in relation to changing marsh Vegetation, American Midl. Nat. 95(2): 313-322.

31

Studies on the Primary Productivity of a Fresh Water Tropical Pond of Madhepura (Kosi-belt), Bihar

S. Kumar, Punam Kumari, Rajnish Kamal and Subrnodeo
Department of Zoology,
P.Sc. College,
Madhepura, Bihar

Abstract

Seasonal changes in the primary productivity of a permanent fresh water fishery ponds of Madhepura Singheswar (Kosi-Beit) was investigated during the different seasons of the year 1998-99. Light intensity and temperature were found to had direct effect on the productivity and light was the important factor limiting productivity. The maximum value of net primary production observed in summer season, in the month of April, was found to be 3.80 gc/m^2/day, which was 7 to 8 times greater than the minimum value of 0.52 gc/m^2/day observed in the month of September 1998. Summer values of net primary production ranged from 1.10 to 3.80 gc/m2/day. Respiration expressed as percentage of gross production amounted to 21.6 to 74.2 and the net gross ratio varied from 0.25 to 0.71.

Introduction

The importance of primary productivity studies in an aquatic environment is well realised in view of its vlaue in estimating the productive capacity. It is not only useful for comparing the productivity of different geographical region, but also for the practical value in fish culture programme. Studies in primary productivity of

South Indian ponds have been performed mostly by Sumitra (1971) and restricted to photosynthesis only. While that of North Indian ponds by Khan and Siddiqui (1971), Ahmad and Siddiqui (1991) and Nasar and Munshi (1975) which are mostly done in macrophytes only. But no work of this type has yet been performed on the fresh water ponds of Kosi-Belt except Kumar, S. and S.C. Yadav (1998). Hence, the present paper has been prepared to study the primary productivity of a fresh water tropical pond of Madhepura and Singheshwar of Kosi-Belt of unique type. Showing the maximum and minimum value of net production and the major factors affecting this productivity in different months of the year 1998-99. The particular tropical pond which is eutrophic in nature is situated in the main town of the Madhepura and Singheshwar. The city of Madhepura is located on the north eastern part of the river Kosi with a North latitude of 25°8′ to 26°20′ and 86°50′ to 87°9′ East longitude and area $(1788.5)^2$ Km.

Materials and Methods

For the primary productivity studies, the light dark bottle method was employed as is given by Gardner and Gran (1927) with 24 hrs incubation period. The samples were suspended at 0.5 m depth of water successively up to 1.0 m. After incubation, the dissolved oxygen was fixed in the field and estimated by the unmodified Winkler's method (Ellis *et al.*, 1948). All oxygen values were converted to carbon values by multiplying by the factor 0.375. By totalling the values of a column of water, the photosynthetic rate under one square metre of water surface was determined to be gc/m²/day, as given by Odum (1963). Light penetration was measured by a Secchi disc.

Observation and Results

The monthly values of productivity alongwith the transparency and water temperature data are shown in Table 31.1.

Discussion

It is evident from the table, that in this particular pond, the primary productivity reached a peak, twice during the annual cycle of November and April 1998. The rate of production was comparatively lower during September and October. Goldman and Netzel (1963) reported that the primary productivity values were typically bimodal in the clear lake of California. Sreenivasan (1964) however reported that the maximum production was in April and the minimum in January in Amaravatty reservoir of South India. The present studies reveal that there is seasonal variation in the productivity values. Sometimes, this seasonal variations in primary productivity, may be quite great, maximum value being seven times greater than that of the minimum values in this same type of water body. Vijaya Raghavana (1971) again reported that maximum value was ten times greater than the minimum values in some pond of Southern India. Qasim *et al.* (1969) also reported the same matter in some water bodies of Cochin and respiraton, expressed as percentage of gross production, amounted to 22.5 to 78.2. Broadly the factors influencing productivity are classified under three heads - edaphic, morphometric and climatic. In this pond the temperature has a direct effect on the productivity. The high productivity (3.80 gc/m²/day) in April, may be correlated with the high temperature (34.4°C) of

Table 31.1: Monthly variation in the primary productivity of the pond

Month	*Temp-erature*	*Trans-parency*	*Production (gc/m²/day)*		*N.P. Vs. G.P.*	*Respiration*	
			Net Produ-ction	*Gross Produ-ction*		*gc/m²/ day*	*% of G.P.*
April 1998	34.4	92.0	3.80	4.85	0.71	1.05	21.6
May 1998	33.8	86.4	3.03	4.18	0.70	1.15	27.2
June 1998	34.2	88.5	1.72	4.12	0.41	2.40	58.9
July 1998	33.7	85.4	1.65	3.65	0.45	2.00	54.7
August 1998	30.2	40.5	0.66	2.56	0.25	1.90	74.2
September 1998	32.3	50.0	0.52	1.23	0.42	0.71	51.6
October 1998	22.8	46.0	1.18	2.40	0.49	1.22	50.8
November 1998	20.8	44.0	2.12	3.78	0.56	1.66	43.9
December 1998	18.4	49.7	1.22	2.84	0.42	1.62	57.4
January 1999	16.4	48.5	0.75	2.93	0.25	2.18	73.2
February 1999	21.9	70.0	1.10	2.42	0.45	1.32	54.9
March 1999	26.8	86.0	2.14	3.48	0.61	1.34	31.1

Gross productivity = O_2 produce - O_2 loss due to respiration (LB - DB) in terms of carbon

Net productivity = Gross productivity - Respiration

that period. The productivity declined to 0.52 gc/m2/day with the full in temperature. Hence it is clear from this particular experiment, that climatic condition such as light is a limiting factor influencing productivity. Light intensity and temperature have a direct effect on productivity. This is in confirmity with the findings of Hickman (1973), Sumitra (1971), Haniffa (1979), Haniffa and Pandian (1978), Nasar and Munshi (1979), Dutta *et al.* (1983), Sreenivasan (1991), Ahmad and Siddiqui (1991) and Kumar and Yadava (1998).

Acknowledgement

Authors are highly greatful to Mrs. Dr. S. Kumar and daughter Shalini for their regularly warm accommodation and constant help during research period.

References

Ahmad, Md. Shamim and E.N. Siddiqui (1991). Limnological Studies of two ponds of Darbhanga, National Conference on Aquatic Sciences in India. New Delhi. pp. 8.

Dutta, N.C., Bandopadhyay, B.K. *et al.* (1983). Diurnal Variation of some abiotic factors and plankton production in two fresh water ponds at Calcutta, Environ. and Ecology - 1. 129-133.

Garder, T. and Gran, H.H. (1927). Production of plankton in the Oslo Fjord. Re. app. proc. verb. cons. perm. Int. Explor. Mer. 42: 9-48.

Goldman, C.R. and Wetzel, R.G. (1963). A study of the Primary productivity of cleare lake, California Ecology, 44: 283-294.

Haniffa, M.A. and Pandian, T.J. (1978). Morphometry, primary productivity and Energy flow in a tropical pond, Hydrobiologia. 59:1, 23-48.

Haniffa, M.A. (1979). Studies on monthly variation of macrophytic biomass in pond idumban, Palni, Comparative Physiology and Ecology, Vol. 4, 2: 64-67.

Hickman, M. (1973). The standing crop and primary productivity of the phytoplankton of Abot's pond. North Somerset J. Ecol., 61: 269-287.

Khan, A. and Siddique, A.Q. (1971). Primary production in a tropical fish pond at Aligarh, India. Hydrobiologia. 37: 447-456.

Kumar, S. and S.C. Yadava, (1998). A Thesis Submitted to the B.N. Mandal University. "Ecobiological Studies of rotifera with special reference to population behaviour of fresh water fisheries ponds at Madhepura".

Nasar, S.A.K. and Datta Munshi, J.S. (1975). Studies on the Primary production of a fresh water pond. Japanese Journal of Ecology, 125:1, 21-23.

Odum, E.P. (1963). Ecology, modern, Biol. Ser. 152 pp. Holt Rinchart and Winston, New York and London.

Qasim, S.Z., Wellershous, S. and Bhattathir, P.M.A. (1969). Organic production in a tropical estuary, Proc. Ind. Acad. Sci., 2: 51-94.

Sreenivasan, A. (1963). Primary production in three up land lakes of Madras State, India, Curr. Sci. 32: 130-131.

Sreenivasan, A. (1991). Reservoir limnology in India. DOEN Lakes Research Project, Kodaikanal 624103. National Conference on Aquatic Sciences in India, New Delhi, 1991. pp. 2-3.

Sumitra, V. (1971). Seasonal Variation in the Primary Productivity in three tropical ponds, Hydrobiologia. 38: 395-408.

Vijaya Raghavan, S. (1970). Seasonal events in a natural population of Daphnia Carinata Kind. Proc. Ind. Acad.. of Sciences. 71, 3(6): 193-203.

Vijaya Raghavan, S.(1971). Seasonal Variations in primary productivity in three tropical pond. *Ibid*, 38: 395-408.